# United Nations University Series on Regionalism

Volume 25

The United Nations University Series on Regionalism, launched by UNU-CRIS and Springer, offers a platform for innovative work on (supra-national) regionalism from a global and inter-disciplinary perspective. It includes the World Reports on Regional Integration, published in collaboration with other UN agencies, but it is also open for theoretical, methodological and empirical contributions from academics and policy-makers worldwide.

Book proposals will be reviewed by an International Editorial Board.

The series editors are particularly interested in book proposals dealing with:

- comparative regionalism;
- comparative work on regional organizations;
- inter-regionalism;
- the role of regions in a multi-level governance context;
- the interactions between the UN and the regions;
- the regional dimensions of the reform processes of multilateral institutions;
- the dynamics of cross-border micro-regions and their interactions with supra-national regions;
- methodological issues in regionalism studies.

Accepted book proposals can receive editorial support from UNU-CRIS for the preparation of manuscripts.

Please send book proposals to: pdelombaerde@cris.unu.edu and lvanlangenhove@cris.unu.edu.

Mario Torres Jarrín • Shaun Riordan

# Science Diplomacy, Cyberdiplomacy and Techplomacy in EU-LAC Relations

Springer

Mario Torres Jarrín
Institute of European Studies and
Human Rights
Pontifical University of Salamanca
Salamanca, Spain

Shaun Riordan
European Institute of International Studies
Primusgatan 53
Stockholm, Sweden

ISSN 2214-9848 ISSN 2214-9856 (electronic)
United Nations University Series on Regionalism
ISBN 978-3-031-36867-7 ISBN 978-3-031-36868-4 (eBook)
https://doi.org/10.1007/978-3-031-36868-4



This Springer imprint is published by the registered company Springer Nature Switzerland AG
The registered company address is: Gewerbestrasse 11, 6330 Cham, Switzerland

# Contents

The original version of this book has been revised. The volume number has been corrected throughout the book. The correction to this book can be found at https://doi.org/10.1007/978-3-031-36868-4_11

# Abbreviations

| | |
|---|---|
| AAAS | American Association for the Advancement of Science |
| ALADI | Latin American Integration Association |
| ALCUE NET | Latin American, Caribbean and EU Network on Research and Innovation |
| AUGM | Montevideo Group Association of Universities |
| APT | Advanced Persistent Threat |
| CALC | Latin American and Caribbean Summit |
| CARICOM | Caribbean Community |
| CELAC | Community of Latin American and Caribbean States |
| CERN | European Organization for Nuclear Research |
| CIA | Central Intelligence Agency |
| CRA | Common Research Area |
| CUIB | Latin American University Council |
| CYTED | Latin American Program of Science and Technology for Development |
| DDOS | Distributed Denial of Service |
| ECLAC | Economic Commission for Latin America and the Caribbean |
| ECSC | European Coal and Steel Community |
| EEAS | European External Action Service |
| ERANET-LAC | EU, Latin American and the Caribbean countries' network on joint innovation and research |
| GGE | Group of Government Experts |
| GESDA | Geneva Science and Diplomacy Anticipator |
| IADB | Inter-American Development Bank |
| IAEA | International Atomic Energy Agency |
| ICANN | Internet Corporation for Assigning Names and Numbers |
| ILAFA | Latin American Iron and Steel Association |
| IRA | Internet Research Agency |
| JCPOA | Joint Comprehensive Plan of Action |
| JIRI | Joint Initiative for Research and Innovation |
| LAFTA | Latin American Free Trade Association |

| | |
|---|---|
| LAWS | Lethal Autonomous Weapon Systems |
| MERCOSUR | Common Market of the South |
| NISA | New International Security Agenda |
| NMP-DELA | Nanosciences, Nanotechnologies, Materials and New Production Technologies Deployment In Latin American Countries |
| NSA | National Security Agency |
| OAS | Organization of American States |
| PCC | Private Cybersecurity Companies |
| S4D4C | Science for/in Diplomacy for Addressing Global Challenges |
| SDGs | Sustainable Development Goals |
| SDR | Special Drawing Rights |
| SICA | Central American Integration System |
| SOM | Senior Officials Meeting |
| TFEU | Treaty on the Functioning of the European Union |
| UNCTAD | UN Conference on Trade and Development |
| UNOEWG | UN Open-Ended Working Group |

# Chapter 1
# Introduction

This book focuses on three themes. Two are in the title: the relations between the EU and the Latin American and Caribbean regions; and the development of the application of diplomacy to science, cyberspace and technology. The third theme is the existence of an activity called diplomacy with its own practices and practitioners which adds value to the management of international relations. It is this latter theme that has been largely ignored in the literature. There is an extensive academic literature on both the relations between Europe and Latin America and the development of international relations in respect of science. Although initially treated as primarily technical or commercial issues there has been a growing interest in the geopolitical implications of cyberspace and technology (Riordan, 2018, 2019; Barrinha & Renard, 2017; McCarthy, 2018). But science diplomacy, for example has focused almost exclusively on how to develop relations between scientists at international level or the infiltration of scientific issues into international debates. There is often an implication that these international issues would be managed so much better if left only to scientists without the interference of diplomats and politicians. Despite excellent recent studies of the nature of diplomacy and diplomats (Sharp, 2009), there have been limited efforts to integrate these studies into discussions about science or technology. But, as French Prime Minister George is reported to have commented after the First World War, war is too serious to leave to soldiers. Likewise, as COVID-19 should have taught us, science is too serious to leave to scientists and technology is too serious to leave to technicians. At an international level at least, diplomats and diplomacy have a role to play.

This book therefore makes three arguments: that science, cyberspace and technology are increasingly important in international relations; that diplomacy and diplomats have a key role in managing these issues at an international level and that this requires a better understanding of diplomatic practices and practitioners and that developing effective approaches to science, cyber and technology diplomacy can play a key role in reviving the inter-regional relationship between the EU and the



M. Torres Jarrín, S. Riordan, *Science Diplomacy, Cyberdiplomacy and Techplomacy in EU-LAC Relations*, United Nations University Series on Regionalism 25, https://doi.org/10.1007/978-3-031-36868-4_1

Latin American and Caribbean Economic Community (CELAC). In making this argument, the book conceives of the interaction between diplomacy and science, cyberspace and technology running along three axes: agency, process and subject matter. Agency addresses the question of who the key actors in the international management of science, cyberspace and technology are. What are the respective roles of scientists, IT engineers and technologists and how do they relate to the role of diplomats? For example, who qualifies as a science diplomat and what are the criteria? Process looks at the tools and techniques used by diplomats in pursuit of broader diplomatic or foreign policy objectives. In the context of technology, this is often described as digital diplomacy and focuses on how diplomats are able to use digital and other technologies in enhancing their capabilities as diplomats. There is already an extensive literature about digital diplomacy (Bjola and Holmes, 2015) and it will not be a major focus of this book. Subject matter explores the application of diplomacy to science, cyberspace and technology. As suggested above, a major theme of this book will be that there is such a thing as diplomacy which must be taken seriously in its own terms, and which has a major role in the management of problems and disagreements in science, cyberspace and technology at an international level. Underplaying the role of diplomacy and diplomats in, for example science diplomacy can lead to serious misunderstandings about the different natures of science and diplomacy and the implications for policy advocacy (this will become especially important in the discussion of the lessons to be drawn from the COVID-19 pandemic).

The book's approach to European–Latin American relations focuses on the inter-regional relations between the EU and Latin American and Caribbean regional organisations. Inter-regionalism is a multidimensional phenomenon (Baert et al., 2014) developed during the last decades. The regional integration processes have reconfigured international relations between countries, modified global governance and changed the international system generating a new world order. With the Schuman Declaration, 9 May 1950, European regional integration marked the beginning of a new stage in its political history and a new way of doing international relations. Seventy years later, many countries have created regional integration organisations taking as a reference the model of European integration. 1945–1970 is defined as the classic period of regional integration (De Lombaerde & Söderbaum, 2013). Scholars consider this period as the starting point for the history of regional integration and have debated, supported and criticised regional integration organisations. Some academics believe that the age of nation states has come to an end and that it is necessary to re-think countries' foreign policies, redefining allies, prioritizing regions, and selecting new markets for its companies exporting goods and services to other countries. This book does not go to these extremes but focuses on the tortuous efforts of the EU to establish inter-regional relations with Latin America and the Caribbean.

The European Union (EU) and Latin American and Caribbean (LAC) inter-regionalism were developed in parallel with the European and Latin American integration processes. These regional integration processes created regional integration organisations, some with political and geopolitical aims and others with economic and commercial purposes. Today, the world has changed, the geopolitical actors extend beyond the Nation State, and the regional integration organisations seek to be geopolitical actors in their region and on the international scene. At the same time new actors have emerged. These include the major internet and technology companies, whose economic power exceeds the GDP of all but the world's largest economies and whose influence in international affairs is increasingly turning these companies into new geopolitical actors in their own right. Their commercial and economic activities are developed within cyberspace but also embedded within the physical world. In the past countries fought wars over territory, resources and the balance of power. These battles were fought out on land, at sea and most recently in the air. Now new domains of conflict have emerged, both in space and cyberspace. Cyberspace is a domain largely built and led by the major technology companies. Their levels of innovation allow them to develop and obtain technological advances that have modified the way of producing goods and services as well as exchanging information and communicating. Nations States do not have similar levels of knowledge and expertise in cyberspace within their government institutions.

The digitalisation of the economy will force Nation States to develop international norms for cyberspace and other digital technologies, both in terms of security and international governance. Artificial Intelligence, Big Data, Machine Learning or Bio-robotics are only some of the areas that pose threats to the security and economic well-being of our societies if they are not regulated at an international level. Cyberspace and the potential development of a metaverse need international agreements on property rights, enforcement of contracts and taxation if they are going to prove economically viable over the longer term. Governments can regulate these areas unilaterally, and the EU is already doing so in areas like data protection and artificial intelligence. But such regulation within limited geographical areas risks putting economics at a competitive disadvantage if other countries or regions do not follow suit. Effective regulation of these technologies must be global. But effective regulation of these areas also requires engagement with the major Internet and technology companies that ultimately must deliver the regulated activities. In this both the European Union and Latin America/Caribbean share a problem in that neither has major Internet or technology companies of their own. Both must find ways of engaging with major US or, in the case of Latin America/Caribbean, Chinese companies. Science Diplomacy, Cyberdiplomacy and Techplomacy can become instruments that allow European, Latin American and Caribbean governments to develop relations both with other governments and Internet and technology companies. But as the European experience has shown, the development of such relations is more effective at the regional rather than nation state level.

The EU and LAC have signed around 27 association, trade or political and cooperation agreements. The EU has negotiated various agreements at different levels: regional, sub-regional and bilateral. At the sub-regional level agreements exist with the Caribbean region, Central American region (SICA) and MERCOSUR. In bilateral terms, the negotiations with the Andean region failed, leading the EU to conclude bilateral agreements with Colombia and Peru instead. In 2011, the Community of Latin American & Caribbean States (CELAC) was created with a commitment to advance the regional integration process in the LAC area. CELAC is looking for unity in political, economic, social and cultural terms. At the same time, it has since become the EU's official counterpart for its strategic partnership with LAC countries.

According to Hänggi (2020), there are three types of inter-regionalism: (1) Traditional or pure inter-regionalism (relations between regional groupings, for example EU-SICA, EU-MERCOSUR or EU-CARICOM). (2) Trans-regionalism (relationships where states participate in their individual capacity in international political forums or bi-regional summits (such as APEC or EU-LAC Summit). (3) Hybrid inter-regionalism (the relation between regional groupings and single emerging powers (such as the political dialogue between EU-Brazil, EU-Chile and EU-Mexico). But EU-CELAC relations generated a new type of inter-regionalism. Ayuso and Gardini (2015) ask how this new inter-regionalism can be defined. According to Torres Jarrín, EU-CELAC inter-regionalism is a "Multiplayer-Inter-regionalism" (Torres Jarrín, 2018), because both regions have promoted the participation of multiple players from the public and private sectors: politicians, officials, diplomats, members of regional organisations, non-governmental organisations and foundations, businesspeople, unions, academics and scientists. In inter-regionalism terms, EU-LAC relations were developed through the EU-CELAC summit system. In recent years, the EU-CELAC Summits of Heads of State and Government have been frozen. The last summit was held in 2015 due to the political crisis in Venezuela. The EU-CELAC Summits resumed on 17–18 July in Madrid. But nevertheless, the EU-CELAC Joint Initiative on Research and Innovation (JIRI) Meetings remain. These meetings, held at senior official level, emphasise the key role of research and innovation in sustainable development and growth and are committed to the development of a bi-regional innovation agenda. The EU has proposed a Science Diplomacy approach, promoting the creation of a "Common Area for Higher Education, Research and Technology". The last VIII Senior Officials Meeting of the EU-CELAC-JIRI, held on 30 October 2020, sought to focus on inter-regionalism in a science diplomacy context. EU-CELAC has been working jointly on research and innovation collaboration since 2010, when the Joint Initiative on Research and Innovation was established to help advance bi-regional dialogue. Ten years have passed and it is time to evaluate whether this cooperation area (Science Diplomacy) has strengthened the EU-LAC relationship and has contributed to creating a common agenda. Most important, can science diplomacy serve as the axis of a new cooperation area in EU-LAC inter-regionalism, and can this cooperation contribute to promoting the UN Sustainable Development Goals and 2030 Agenda?

In 2010, the concept of Science Diplomacy was popularised by the Royal Society and American Association for the Advancement of Science (AAAS) through its report "New Frontiers in Science Diplomacy" (The Royal Society and AAAS, 2010), which highlighted the universal character of science and its capacity to cross borders and connect societies around the world. At the same time, science diplomacy offers the opportunity to address global challenges. According to the AAAS, there are three types of science diplomacy: science in diplomacy, diplomacy for science and science for diplomacy. Gluckman et al. (2017) proposed a new categorisation based on a government's motivations for its foreign policy: promote national interests, resolve trans-borders problems and address global challenges. For Van Langenhove and Burgelman the older concept of science diplomacy urgently needs to be expanded and to be updated in the wake of the European Commission's announcement that it will re-think its international R&D collaboration and indicated that according to three major Swedish universities (known as "The Stockholm Trio") science diplomacy should be seen as a positive trademark for Europe (Van Langenhove & Burgelamn, 2021). In 2016, the UNESCO Report on Science suggested that the world is adopting new development strategies based on science, technology and innovation. Countries try incorporating these three areas in their national research systems to improve their competitiveness and brand themselves as an "innovator country". In 2018, the European Commission published "Europe's future; Open innovation, Open Science, open to the world: Reflections of the research, innovation and Science Policy Experts (RISE) High Level Group". In both cases, it mentioned the importance of international cooperation. Through international cooperation countries can implement a "soft power" whose instrument is Science Diplomacy. Despite this work, much confusion remains around the concept of Science Diplomacy (Riordan & Torres Jarrín, 2020a), in part because of the non-involvement of diplomats or scholars of international relations and diplomacy in these discussions. This book will aim both to place the concept of science diplomacy on a firmer intellectual basis and to explore how science diplomacy, once more rigorously established, can serve to reinforce inter-regional cooperation between the EU and CELAC. In doing so it will look not only at science diplomacy narrowly defined, but also at related concepts like cyberdiplomacy (Riordan & Torres Jarrín, 2020b) and Techplomacy (Riordan & Torres Jarrín, 2020c), exploring their relevance to CELAC countries and CELAC's relations with the EU. For the purposes of this book, science diplomacy focuses on the interaction between science and diplomacy, whether through agency, process or domain. Put another way: who is doing the diplomacy, what techniques are they adopting to do the diplomacy and what is the subject matter to which diplomacy is being applied? Science engages with diplomacy along all three axes. At first, the concept was limited to international exchanges between scientists (agency). But as the subject has matured it has engaged more effectively with international relations and international security, involving all three of the axes of diplomacy. The growing importance of science and technology in international relations has meant diplomats having to tackle scientific issues as scientists have been forced to engage with diplomacy. Three areas have driven these developments: the development of digital technologies and artificial intelligence, and their implications for global governance; the climate change debate; and COVID-19. All these areas are relevant to CELAC countries, EU countries and the relations between them.

As in other areas, COVID-19 has accelerated pre-existing trends in diplomacy, forcing diplomats online and to confront scientific and technical issues as core parts of foreign policy agendas. It is not simply a question of scientists taking on the role of diplomats: many of the issues arising from COVID-19 are political or geopolitical rather than scientific. Some of the scientific issues (e.g. the phylogenesis of COVID-19) are being contaminated by geopolitics. In these cases, the challenge for science diplomacy is to create neutral platforms where scientists can continue their investigations unaffected by the political blame game. But the impact of COVID-19 extends beyond the purely medical. By forcing more activities online, and a greater use of technology, the virus has increased our societies' cyber vulnerabilities. The development of vaccines against the virus generates a new "vaccine diplomacy" as the international community develops protocols around the distribution and use of the vaccines. Even once the virus is over, the international community will need to negotiate new rules and protocols for the global movement of goods and people between societies newly conscious of their vulnerability to pandemic disease. In all these areas of science and diplomacy, scientists and diplomats, will have to find new ways of engaging with and working with each other.

Similar diplomatic and scientific challenges exist in other areas of science, and in relation to the UN's Sustainable Development Goals (SDGs) and its 2030 Agenda. The greater dependency on digital technologies highlights the related areas of Cyberdiplomacy and Techplomacy. Cyberdiplomacy explores the application of diplomacy to the problems arising in cyberspace, whether related to Internet governance or cybersecurity. These problems are as much political and geopolitical as they are technical. Techplomacy recognises the importance of the major Internet companies, including hardware and software providers, search engines or social media platforms, to twenty-first century global governance. Treating them as geopolitical actors, it seeks to engage them constructively in the key Internet governance and cybersecurity debates. As far as technology contributes to the SDGs and Agenda 2030, Cyberdiplomacy and Techplomacy become essential adjuncts to science diplomacy. The countries of CELAC and the EU both have an interest in cyberspace regulated by internationally agreed norms, and both must confront cyberspace without major technology giants of their own. There is much that should form the basis for close inter-regional collaboration.

This book investigates the impact of Science Diplomacy, Cyberdiplomacy and Techplomacy in the relations between the European Union and Latin America and the Caribbean, and how collaboration in these areas could reinforce its inter-regionalism, strengthen society resilience in both CELAC and EU members and contribute to realising the SDGs and Agenda 2030 objectives. Chapter 2 explores the history of EU-CELAC inter-regionalism. It seeks to contextualise this bi-regional collaboration in the spheres of cooperation in science, technology and innovation. It also analyses the problems that have hindered the development of EU-CELAC collaboration in the past and considers the extent to which science diplomacy has helped overcome these problems in the past and can in the future drive a more effective inter-regional cooperation agenda.

Chapter 3 begins a section of four chapters that seek to place science and technology in a broader diplomatic and geopolitical context. Chapter 3 seeks to provide a solid conceptual basis for science diplomacy, setting it within the context of international relations and diplomacy studies. It explores the three axes along which science and diplomacy engage (agency, process and subject matter), drawing on historical and current examples of all three. It examines the relations between scientists and diplomats and how these are changing in the twenty-first century. It sets science diplomacy in the context of the twenty-first century geopolitical and scientific environments, exploring how it is affected by changes in both, including in the relative international balance of power.

Chapter 4 turns to the application of diplomacy to cyberspace and digital technologies. It builds on the theoretical framework constructed in the previous chapter to situate cyberdiplomacy in relation to science diplomacy and the broader international and geopolitical environment. It will argue that the major problems in cyberspace are political and geopolitical and therefore not susceptible to technical solutions. To the extent that these problems are international it argues that governments need to develop a new cyberdiplomacy to manage these issues at a global level. It explores the conflict of cyber agendas between Internet governance and cybersecurity and the challenges of building international norms of behaviour in cyberspace.

Chapter 5 extends the discussion into Techplomacy. It argues that technology is defining the geopolitical environment in three senses: technology is shaping the geopolitical environment, both by providing the tools of geopolitical competition and through the geographical concentration of the major technology companies; technology itself is becoming a battleground for geopolitical rivalries at a global level; and technology companies themselves are increasingly behaving like geopolitical actors. It investigates the challenges posed for governments in engaging with the major technology companies and evaluates the decision of the Danish Government to appoint an ambassador to the technology sector. It compares the situations of both CELAC and the EU in relation to major technology and Internet companies, noting the similarities and differences and the areas of possible collaboration. It suggests that Techplomacy could prove a major and productive area for EU-CELAC collaboration, especially in light of the geographical concentration of major technology companies in the USA and China.

This section concludes with Chap. 6 which seeks to draw conclusions from the experiences of the climate change debate and managing the COVID-19 pandemic. The chapter analyses the performance of science diplomacy in relation to the COVID-19 pandemic, and in particular how the pandemic highlighted the differences between science and policy. It suggests that the science outperformed the diplomacy and that scientists must be conscious of the difference between acting as scientists and acting as diplomats or policy advisors. It looks at the lessons that can be learnt and at the challenges for science diplomacy in relation to COVID-19 going forward. It concludes by looking at the lessons for science diplomacy that can be learnt from the experience of negotiating climate change agreements.

Chapter 7 returns the focus to EU-LAC relations. It looks at science diplomacy in relation to the SDGs and Agenda 2030. It also relates EU-CELAC interregionalism to the pursuit of the objectives set by the UN through both the SDG and Agenda 2030. It argues that multistakeholder diplomacy is central to achieving the objectives set out by the United Nations. Finally, it examines how this multistakeholder diplomacy can drive and be driven by EU-CELAC cooperation.

Chapter 8 focuses on science diplomacy as a key area of cooperation in science and technology between EU-CELAC. It looks at existing bi-regional projects and how these can be developed further in the future. The EU-LAC inter-regionalism has fostered a bi-regional cooperation based on science and Research & Development (R&D). It is time to evaluate how collaboration in science and technology between the EU and CELAC can advance towards an EU-CELAC science diplomacy, who will be the direct beneficiaries in the societies in both regions and the indirect beneficiaries in a global society. It argues that the development of an EU-CELAC science diplomacy can contribute a better global governance and rebalance the international system.

Chapter 9 returns to the themes of Cyberdiplomacy and Techplomacy and explores the scope for collaboration between EU-CELAC in these areas. It looks at the importance of the development of the digital economy and relates it to existing EU-CELAC joint programmes. It suggests ways in which CELAC and the EU could collaborate, both in promoting a ruled-based cyberspace and in strengthening digital resilience. It also explores the scope for inter-regional collaboration in these areas in relation to managing relations with internet giants and generating local competitors. It concludes by evaluating the appointment of Techambassadors by CELAC and the EU, either separately or jointly.

Chapter 10 seeks to draw together the various themes running through the book. It argues that science and digital technologies, including cyberspace, are increasingly important themes in the international agenda. It is not simply a question of making effective use of digital technologies in pursuing broader diplomatic and foreign policy objectives. Digital technologies are increasingly generating problems whose management at an international level needs new diplomatic approaches. Science and digital technologies are becoming new domains of diplomatic activity. Common features in the situations in which CELAC and the EU find themselves in relation to science and technology, in particular, the absence of major technology and Internet companies in either region, suggest that they enjoy shared interests upon which future inter-regional collaboration could be built. An EU-CELAC inter-regionalism centred on science and technology could create a new inter-regionalism 4.0 for a new global governance 4.0.

## References

Ayuso, A., & Gardini, G. L. (2015). EU-Latin America and Caribbean inter-regional relations: Complexity and change. *Atlantic Future Project*. Retrieved December 12, 2021, from https://www.cidob.org/en/publications/publication_series/project_papers/atlantic_future_papers/scientific_paper/eu_latin_america_and_caribbean_inter_regional_relations_complexity_and_change

Baert, F., Scaramagli, T., & Söderbaum, F. (Eds.). (2014). Intersecting interregionalism. Regions, global governance and the EU. United Nations University Series on regionalism 7. New York. Springer.

Barrinha, A., & Renard, T. (2017). Cyberdiplomacy: The making of an international society in the digital age. *Global Affairs, 3*, 353–364.

Bjola, C., & Holmes, M. (2015). *Digital diplomacy: Theory and practice*. Routledge.

De Lombaerde, P., & Söderbaum, F. (Eds.). (2013). *Regionalism* (Vol. 1–4). Sage.

Gluckman, P. D., Turekian, V. C., Grimes, R. W., & Kishi, T. (2017). Science Diplomacy: A pragmatic perspective from the inside. *Science and Diplomacy*, Vol. 6 N 4. Retrieved December 12, 2021, from https://www.sciencediplomacy.org/article/2018/pragmatic-perspective

Hänggi, H. (2020). *Interregionalism: Empirical and theoretical perspectives*. Paper prepared for the Workshop "Dollars, Democracy and Trade. External Influence on Economic Integration in the Americas". Los Angeles. The Pacific Council on International Policy.

McCarthy, D. (Ed.). (2018). *Technology and world politics: An introduction*. Routledge.

Riordan, S. (2018). The geopolitics of cyberspace: A diplomatic perspective. *Brill Research Perspectives in Diplomacy and Foreign Policy, 3*(3), 1–84.

Riordan, S. (2019). *Cyberdiplomacy: Managing security and governance online*. Polity.

Riordan, S., & Torres Jarrín, M. (2020a). *Cyberdiplomacy, global policy perspective report*. European Institute of International Studies. Salamanca-Stockholm. 2020. Retrieved December 12, 2021, from https://www.ieeiweb.eu/wp-content/uploads/2020/01/Cyberdiplomacy_enero.pdf

Riordan, S., & Torres Jarrín, M. (2020b). *Techplomacy and the tech ambassador*. Global Policy Perspective Report, European Institute of International Studies. Salamanca-Stockholm. 2020. Retrieved December 12, 2021, from https://www.ieeiweb.eu/wp-content/uploads/2020/01/Techplomacy-and-the-Tech-Ambassador.pdf

Riordan, S., & Torres Jarrín, M. (2020c). *Science diplomacy, global policy perspective report*. European Institute of International Studies. Salamanca-Stockholm. 2020. Retrieved December 12, 2021, from https://www.ieeiweb.eu/wp-content/uploads/2020/01/Science-Diplomacy.pdf

Sharp, P. (2009). *Diplomatic theory of international relations*. Cambridge University Press.

Torres Jarrín, M. (2018). EU-CELAC: A multiplayer interregionalism. Redefining the Atlantic area. In G. L. Gardini, S. Koschut, & A. Falke (Eds.), *Interregionalism and the Americas* (pp. 37–54). Lexington Books.

The Royal Society and AAAS. (2010). *New frontiers in science diplomacy: Navigating the changing balance of power*. Retrieved December 12, 2021, from https://royalsociety.org/~/media/Royal_Society_Content/policy/publications/2010/4294969468.pd

Van Langenhove, L., & Burgelamn, J. C. (2021). Viewpoint: Science diplomacy needs a refresh to meet contemporary European needs. *Science Business*. Retrieved December 12, 2021, from https://sciencebusiness.net/viewpoint/viewpoint-science-diplomacy-needs-refresh-meet-contemporary-european-needs

# Chapter 2
# European Union-Latin America and the Caribbean Inter-regionalism History

**Abstract** The chapter explains the European Union-Latin American and Caribbean (EU-LAC) inter-regionalism history and secondly, contextualises the bi-regional cooperation agenda in science, technology and innovation. At the same time, it analyses the problems that have hindered EU-CELAC collaboration in the past and considers to what extent science diplomacy has helped in the past and can further help in the future, in mitigating and overcoming the problems before mentioned.

**Keywords** Regional integration · Inter-regionalism · Science · Technology and innovation cooperation · EU-LAC · OAS and EU-CELAC

## 2.1 European Union-Latin America and the Caribbean Relation in the International Context

At the time of writing this book (2022–2023) the international system is undergoing radical change. New world powers and new geopolitical actors have emerged changing the economic and political axes around international relations revolve. The Pacific axis of regional power has replaced the Atlantic axis as the centre of global power. Tensions surround the hegemonic leadership of this region which will decide the future of the Indo-Pacific. The current world order is unrecognisable compared to that in which the inter-regionalism between Europeans, Latin Americans and the Caribbean began.

The following section identifies the decade of the 1950s as the beginning of the regional integration processes that would create the first integration organisations, some with a sub-regional scope and others with a regional and/or continental nature. The creation of the European Coal and Steel Community (CECA) through the Paris Treaty of 1951, which later became the European Economic Community through the Rome Treaties of 1957, will be the pillars of European construction and ultimately what we know today as the European Union, according to the Maastricht Treaty of 1993. Almost parallel to European integration, the integration process of

 
M. Torres Jarrín, S. Riordan, *Science Diplomacy, Cyberdiplomacy and Techplomacy in EU-LAC Relations*, United Nations University Series on Regionalism 25, https://doi.org/10.1007/978-3-031-36868-4_2

the countries of Latin America and the Caribbean is developed: the Organization of Central American States (El Salvador Agreement 1951) and the Andean Community (Cartagena Agreement 1969). In 1964, the visit of a delegation from the European Parliament to some Latin American countries will result in the "Martino Report", which will serve as the basis for developing inter-regionalism between the European Union and Latin America and Caribbean countries. In the 1960s, the international system was exclusively made up of Nation States, which were the only international actors and the only actors with the power to create and implement development policies in countries. Therefore, economic and social developments, as well as representation at the international level, were the exclusive competence of the Nation States.

The 1960s is also the decade when regional integration organisations began to have power over development policies and to intervene in foreign policy matters. Some of them have legal personalities and are subjects of international law, which makes them geopolitical actors on the international stage. As of 01 March 2022, according to the World Trade Organization, there are 354 regional trade agreements, many of them developed under regional integration schemes or organisations. In the same year, the world experienced two great events that have completely changed the international balance: the global consequences caused by the COVID-19 pandemic and Russia's invasion of Ukraine. The Russia-Ukraine war has divided countries into two groups: countries that condemn the invasion and those that refrain from condemning it. Russia has managed to make new alliances, including with China and India, three powers vying for leadership in the Asia-Pacific region. Orient and West seem to continue fighting for world hegemony. But there is a big difference to previous centuries: before the fighting was for the control of a determined region or area between a small group of powers; now there are many powers and many regions to control and dominate.

After the Second World War a new world order was established, dominated by US power and its leadership in the West and the rest of the world. It is true that its leadership had as a rival the Union of Soviet Socialist Republics (USSR). The confrontation between these two powers was not only limited to a struggle for economic, military, and political power. It was an ideological struggle that divided the world into two parts, the capitalist countries or those within the capitalist sphere, and communist countries or those within the communist sphere. After the fall of the Berlin Wall in 1989 and successively the implosion and dissolution of the USSR between 1990 and 1991, world hegemony became exclusive to the USA. By that time, European countries had recovered from Second World War and their economies were booming, their trade expanding, and their influence slowly growing on the international stage.

During the 1990s, the USA was the main trading partner, the largest investor, and the largest contributor of official development assistance to most countries worldwide. This was also true for the countries of Latin America and the Caribbean. However, as we have mentioned before, the European Economic Community, which during the 1990s will become the European Union, takes the place of the USA, while the USA focuses all its attention on the Asia-Pacific region. The following decade, 2000–2010, the EU consolidated itself as the first partner in terms of trade, development cooperation and investment in the Latin American and Caribbean regions, while the USA divided its interest between the Asia-Pacific and Middle East regions. The decade 2010–2020 is also the decade in which Latin America and the Caribbean focus their interest on the Asia-Pacific region, and China begins to take the place of the European Union within the region.

The period 2020–2023 is a transition where a new world order is being configured. The dispute over commercial leadership and political-military hegemony between the great and medium powers will define the future world order. In the reconfiguration of this new world order, the regional blocs and organisations will play a determining role. Thus, we can see how the G7, G20 and NATO seek to maintain the leadership of the West in international affairs. While the Shanghai Cooperation Organization, Eurasia Economic Union or initiatives such as BRICS and One Belt, One Road seek to expand and consolidate the power of Asian countries, mainly the power of China, Russia and India. However, it is important to point out that these initiatives also involve both European and Latin American and Caribbean countries. Consequently, they have direct repercussions on the inter-regionalism developed between the European Union and the Community of Latin American and Caribbean States (EU-CELAC).

## 2.2 European Union-Latin America and the Caribbean Inter-regionalism History

Since the beginning of European integration, European countries considered Latin American countries as their natural allies in global governance affairs. In the early 1960s, the European institutions decided to establish diplomatic relations with Latin American countries with two main objectives:

1. Explore trade opportunities between European and Latin American countries (in particular with the South American countries).
2. Coordinate common positions for the First United Nations Conference on Trade and Development (UNCTAD), which was held in Geneva in 1964.

It should be noted that the UNCTAD marked a new stage in the history of international relations because it represented the first efforts to create an international forum between highly industrialised and developing countries, as well as the recognition of the validity and urgency of their claims for a fairer and more equitable ordering of international trade in terms of development (ECLAC, 1971). At the same time, it represented a change in the reconfiguration of the international system, as it was the first time that a group of countries acted through a regional integration organisation, in the European case, the European Economic Community (EEC). Therefore, the UNCTAD served to facilitate trade and promote inter-regionalism.

In 1962, the EEC transmitted a report to the Foreign Affairs Ministers Council whose conclusions included creating a Latin American Program. Until this moment in the EEC, in international cooperation policy terms, Latin American and Asian countries were integrated into the same regional programme. In March 1963, the Trade Commission of the European Parliament decided to send a special delegation to visit Argentina, Brazil, Colombia, Chile, Peru and Uruguay. The parliamentary delegation had bilateral meetings with the governments and a work meeting with the Latin American Free Trade Association (LAFTA), one of the first regional integration organisations created in February 1960 through the Montevideo Treaty (in the 1980s it was transformed into Latin American Integration Association-ALADI). EEC and LAFTA issued a joint statement where both parties expressed their wishes to establish formal and permanent relations (LAFTA, 1964), setting as goals to promote regional integration and encourage international trade (Gómez Hernández, 2020).

The European Parliament delegation was led by its president Gaetano Martino, who elaborated the so-called "Martino Report". The report was presented on 24 November 1964 and identified as Latin American partners for the European Coal and Steel Community (ECSC): "Latin American Institute of Iron and Steel (ILAFA), Economic Commission for Latin America and the Caribbean (ECLAC) and, the Special Commission Coordination for Latin America (CECLA)" (IRELA, 1996).

The United Nations Economic Commission for Latin America and the Caribbean, known as ECLAC, was established in 1948 as a United Nations regional commission to encourage economic cooperation among the Latin American countries. In 1984, a resolution was passed to include the countries of the Caribbean in the name. ECLAC has its headquarters in Santiago, Chile and is one of the five regional commissions of the United Nations. In 1951, it opened a sub-regional headquarters in Mexico City, which serves the needs of the Central American sub-region. In 1966, it established another headquarters in the Caribbean sub-region, in Port of Spain, Trinidad and Tobago. In addition, ECLAC maintains country offices in Buenos Aires, Brasilia, Montevideo and Bogota, as well as a liaison office in Washington DC. The ECLAC has Member States and Associate Members: the 33 countries of Latin America and the Caribbean, together with several Asian, European and North American nations that have historical, economic and cultural ties with the regions.

ILAFA was created in 1959 on the initiative of ECLAC, with the idea of replicating the experience of the European Coal and Steel Community (ECSC). ILAFA opened its headquarters in Santiago de Chile. The regional organisation changed its name in 2011 due to a series of initiatives that renewed its focus and field of activities. The ILAFA aims were strengthening the value of regional integration, technological innovation, human resources excellence, corporate responsibility, social and environmental sustainability.

In 1964, in the Organization of American States framework, the III Meeting of the Inter-American Economic and Social Council, which also coincided with the Economy Ministers Meeting of 19 Latin American countries, was held. In these meetings, the countries decided to formally establish the Special Commission for

Latin America (CECLA) to coordinate economic and trade affairs at the regional level.

There have been four periods in EU-LAC relations (Torres Jarrín, 2018):

- First period (1964–1991): From Martino's Report to European Economic Community-Rio Group meetings.
- Second period (1991–2009): EEC-Rio Group Institutionalized Ministerial Meetings System.
- Third period (1999–2013): EU-LAC Summit of Head of States and Government System.
- Fourth period (2013-present): EU-CELAC Summits.

As we can observe, over the decades European and Latin American relations have moved from an informal and sporadic relationship to an institutionalised permanent relationship. While it is true both regions have been able to consolidate a strategic partnership, the parties have created an inter-regionalism acquis through the bi-regional dialogues and treaties. The dialogues are realised at multi-level, from senior officials to civil society organisations members.

In 2024 it will be 50 years since the Martino Report. In inter-regionalism terms, relations between European, Latin American and Caribbean countries have developed within the EU-CELAC summits system. CELAC (Community of Latin American and Caribbean States) is heir to the Rio Group, which was created based on the "Contadora Group" and the "Contadora Support Group". Both were initiatives by Latin American foreign ministers to deal with the Central American crisis, caused by armed conflicts in El Salvador, Nicaragua and Guatemala. With stabilisation and peace achieved in Central America, the Latin American countries decided to create the Rio Group as a permanent regional mechanism for consultation and concertation. CELAC was created on 23 February 2010 at the Latin American and Caribbean Unity Summit, held in Playa del Carmen, Mexico. The summit was the joint celebration of the XXI Rio Summit and the II Latin American and Caribbean Summit on Integration and Development (CALC).

The I CALC was a summit of heads of Latin American and Caribbean countries held on 16 and 17 December 2008 in Costa do Sauipe, Brazil. The summit was an initiative of the President of Brazil, Luis Inácio Lula da Silva, with the goal of building cooperation mechanisms with greater autonomy from the USA. The majority of presidents attended the summit, with the exception of the President of Colombia, Álvaro Uribe and the President of Peru, Alan García. The meeting finished with the signing of the Bahia Declaration, a common agenda establishing the following priorities affairs: cooperation between mechanism of regional and sub-regional integration, the global financial crisis, energy, infrastructures, social development and eradication of hunger and poverty, food security, sustainable development, natural disasters, human rights, promotion, migration, South-South cooperation and Latin America and Caribbean protection (Bahía Declaration, 2008).

At the II CALC, the Cancun Declaration was signed which also defined issues to be addressed in its regional agenda and external relations: economy and trade, energy, physical infrastructure integration, science and technology, development and social cohesion, migration, sustainable development, climate change, human rights, security issues, South-South cooperation and drugs (Cancun Declaration, 2010). At the III CALC Summit, held on 2 and 3 December 2011 in Caracas, the organisation was created. It should be noted that although the Caracas Declaration, 2011 is usually mentioned as the founding document of CELAC this is not true. CELAC is not an organisation with legal personality, it was simply a political declaration of intentions, in which CELAC is the forum or political space where regional and extra-regional issues are debated. CELAC acts as a spokesman for the Community with other countries and regional organisations such as the European Union, The Cooperation Council for Arab States of the Gulf, China, the Russian Federation, the Republic of Korea, Turkey and Japan.

In January 2013, the VII EU-LAC Summit/I EU/CELAC summit took place in Santiago (Chile). It was proposed to organize EU-CELAC relations through a system of thematic summits at different levels: governmental, private sector, academia, civil society etc.

EU-CELAC relations have generated a new type of inter-regionalism. According to Hänggi (2000) there are different types of inter-regionalism:

1. Traditional or pure inter-regionalism: This inter-regionalism is developed by regional integration groups, e.g. EU-SICA, EU-MERCOSUR OR EU-CARICOM).
2. Trans-regionalism: Defined as relationships where states participate in an individual capacity in international political forums or bi-regional summits, e.g. APEC or EU-LAC Summits.
3. Hybrid inter-regionalism: Where relations between regional integration groupings and single emerging powers, e.g. the political dialogues between EU-Brazil, EU-Chile and EU-Mexico.

EU-CELAC relations do not fit any of these categories. How do we define the EU-CELAC inter-regionalism? (Gardini & Ayuso, 2015). According to Torres Jarrín, the EU-CELAC inter-regionalism is a "Multiplayer-Inter-regionalism" because the inter-regionalism developed by EU-CELAC countries has promoted the participation of multiple players in the bi-regional dialogue process, actors from the public, private, academia and civil society sectors: politicians, officials, diplomats, businesspeople, members of international organisation, members of governmental and non-governmental organisations, international foundations, unions, academics, scientists, students and so on (Torres Jarrín, 2018).

We can differentiate two stages according to the nature of the participants in the EU-LAC relations. From Martino's Report (1963) to I EU-LAC Summit of Head of State and Government (1999) the participants were government officials: politicians, diplomats and senior officials. The I EU-CELAC Summit (2013) brought together members of the private sector, academia, civil society, etc. These new players have created a bi-regional summit system, which is structured around the EU-CELAC biennial thematic summits: EU-LAC Trade Union, EU-LAC Organized Civil Society, EU-Civil Society Forum, CELAC-EU Youth Days, EUROLAT (Euro-Latin American Parliamentary Assembly), Mayors Forum, EU-CELAC Academic

Summit, Senior Official Meeting on Science and Technology (JIRI), EU-CELAC Foreign Ministers and EU-CELAC Business Summit. The EU-CELAC Action Plan approved in Brussels 2015 identified 10 key cooperation areas, the first area indicated was "Science, research, innovation and technology".

In 2015, the II EU-CELAC Summit of Head of State and Government was held in Brussels. This meeting was the last time that all the leaders met before 2023. There were many factors that complicated the summit celebration, mainly, the differences about how to solve the political crisis in Venezuela. CELAC suffered a double fracture. Firstly, the internal fracture consisted in the division of the Latin American and the Caribbean countries. They were divided into two groups: countries that supported President Nicolas Maduro and considered his government legitimate and the other group of countries that questioned the legitimacy of the Venezuelan President and believed that the democratic order had been broken in Venezuela. Secondly, the external fracture in its relations with the EU. The EU urged Venezuela to restore democratic order. For European countries reconciliation in Venezuela was of the utmost importance both for EU and for countries in the region (European Council, 2016).

On 12 December 2015, a new global agreement on climate change was reached, called the Paris Agreement representing a balanced outcome with an action plan to limit global warming. European, Latin American and Caribbean countries signed the agreement. Climate change was a common point in the bi-regional agenda EU-CELAC; "We commit to work, within the framework of our strategic bi-regional association, in an inclusive manner and on equal terms for both regions, to achieve a legally binding, global climate agreement that is needed to keep global warming below 2 °C and to pursue efforts to limit it to 1.5 °C" (European Council, 2015).

Interestingly, in 2015 the EU-LAC strategic partnership demonstrated its maximum effectiveness and diplomatic power by securing the approval of the Paris Agreement with its votes. But as we have mentioned before, the political crisis in Venezuela overshadowed all kinds of positive results or bi-regional alliance benefits.

Just as UNCTAD was the first international forum where European, Latin American and Caribbean countries acted together to try to create international norms, the Paris Agreement represents the last effort at multilateral level by both regions to promote a binding agreement to combat climate change, adapt to its effects and to establish international standards that allow us to have a better world through sustainable development.

The experience of industrialised countries shows that economic growth with sustainable development is possible only if social inclusion policies are implemented. The success of social inclusion policies depends on the generation of scientific-technological-social environments favourable to innovation. If in the twentieth century, the main arguments for reducing the social gap focused on promoting industrialisation, trade and foreign direct investment policies around the world, today in the twenty-first century, the world faces a double gap, a social gap and a digital gap.

Technological advances have modified the ways of design, production and marketing of goods and services. For this reason, it is necessary to elaborate and implement strategies and action plans on Science Diplomacy, Cyberdiplomacy and Techplomacy. These three areas in themselves represent three potential cooperation areas between EU-LAC countries. The EU-CELAC inter-regionalism developed during the last decades leaves us a bi-regional acquis in many thematic areas related to the three areas mentioned above. But it is essential to design an EU-LAC digital strategy focused on two lines of external action: an inter-regionalism 4.0 and a governance 4.0.

## 2.3 Regional Integration: Geographical Context in Europe, Latin American and the Caribbean

When we are thinking about regional integration as a stage within international relations history, usually we take as a reference the European regional integration process. We place the Schuman Declaration as the foundation stone that originates the European Coal and Steel Community (ECSC) created in 1951 and then the European Economic Community (EEC) established in 1957. Although often reference is also made to the BENELUX Union created in 1944 as a previous experience in integration.

The Schuman Declaration was a proposal made by French foreign minister Robert Schuman on 9 May 1950 that included placing French and West German production of coal and steel under a single authority to prevent future conflict between both countries, among its member states, to pacify the continent, strengthen France-West Germany relations through gradual political integration which would be achieved by creating common interests.

From the beginning, we can observe in the European integration experience a "triple common concept": interests, policies and single authorities (supranational institutions). On the basis of these common points, the European countries have built their regional integration. Concepts mentioned by Latin American and the Caribbean countries in their many regional and sub-regional integration treaties but never considered as a sine qua non condition to lay the foundation for their regional integration processes.

De Lombaerde and Söderbaum (2013) describe the 1945–1970 period as a "regional integration classic period". Many scholars take this period as the point starting of regional integration history and they focus their research, theories and hypotheses on it: Mitrany (1975), Viner (1950), Haas (1958), Mundell (1961), Cantori and Spiegel (1970), Nye (1971), Schmitter (1970), Balassa (1961) or Puchala (1968). Each one of these authors from different perspectives and disciplines has studied the historical evolution of the regional integration processes and their impact on the international system.

Any initiatives on regional integration before the twentieth century have been considered as mere "dreams of unification" or "utopia". However, the history of regional integration was not born in the twentieth century. In the earliest stages of the creation of the European Nation-States and the American Republics, between the seventeenth and twentieth centuries, there were regional integration projects in both continents. This period is called "integrationist acquis".

To understand the regional integration process in Europe, Latin America and the Caribbean regions it is necessary to clarify some concepts such as "region" or "sub-region". According to Russett (1967) to speak of a region, it must meet three conditions: geographical proximity, economic interdependency and homogeneity (history, languages, religion etc.). For a region is a geographical territory sharing a group of countries whose foreign policies are interrelated.

Hettne indicates that from an empirical observation of the regional integration processes around the world, a region is defined as a group of countries that share a political project (Hettne et al., 2002). But they clarify that traditionally a region was considered as a group of countries with geographical proximity and a certain degree of mutual dependence. However, for Fawcett geographical proximity is not enough to define a region. He considers it indispensable to share common elements such as languages, culture, religion, political and economic systems (Fawcett, 2005).

It is not enough to say that a group of countries that decides to begin a regional integration process is a new region in political terms. In that sense, a region is a set of states that seek to transcend the central state (Hettne & Söderbaum, 2002).

Another definition of region is that regions are the creation of a new political power whose purpose is to extend the power of states beyond its borders through strategic actions (Katzenstein, 2005).

Farrell (2005) mentions that a region is not a static form, rather a dynamic form, open to change and adaptation, its building and reconstruction are part of a global transformation process.

Academia has failed to define sub-region. However, in integration in both the European and Latin American and the Caribbean cases there are sub-regional processes.

In the European integration case, we can find 14 sub-regional processes:

1. Benelux Union (1944)
2. Western European Union (1948)
3. Organization for European Economic Co-operation (1948)
4. European Coal and Steel Community (1951)
5. European Economic Community (1957)
6. European Atomic Energy Community (1957)
7. European Free Trade Association (1960)
8. Central European Initiative (1989)
9. Commonwealth of Independent States (1991)
10. Central European Free Trade Agreement (1992)
11. Organization of the Black Sea Economic Cooperation (1992)
12. European Economic Area (1994)
13. Baltic Free Trade Agreement (1994)
14. Eurasian Economic Union (2014)

There are other sub-regional initiatives and processes such as the Cooperation Processes of the South-East Europe (1996), the Adriatic-Ionian Sea Initiative (2000), Danube Cooperation (2002) and the Regional Cooperation Council (2008). Everyone of these treaties has served to promote integration processes.

Cottey indicates that since 1990 there has been a proliferation of "new sub-regional groups", sub-regional integration processes that emerged from Eastern and Central Europe, the Balkans, the Mediterranean and countries of the former Soviet Union (Cottey, 2009).

In the Latin American and the Caribbean case, we can find 14 sub-regional processes:

1. Organization of Central American States (1951)
2. Latin American Free Trade Association (1960)
3. Central American Common Market (1960)
4. Caribbean Free Trade Association (1965)
5. Eastern Caribbean Common Market (1968)
6. Andean Subregional Integration Agreement (1969)
7. Latin American Economic System (1975)
8. Southern Common Market (1981)
9. Association of Caribbean States (1994)
10. South American Community of Nations (2004)
11. Bolivarian Alliance for the Peoples of our America (2006)
12. Community of Latin American and Caribbean States (2010)
13. Pacific Alliance (2012)

These regional integration initiatives and projects show the resilience of regionalism in the Latin American and Caribbean region (Rivarola Puntigliano & Briceño-Ruíz, 2013; Briceño-Ruíz & Rivarola Puntigliano, 2021).

Curiously, the two "regions" of our study are geographically subcontinents or a sub-region if we consider the region in relation to continent. Geographically it is incorrect. Europe which is often defined as a continent is a subcontinent of Asia, really it is Eurasia. In the Latin America and the Caribbean case, it is a subcontinent of the American continent. The only exception that we can cite in continental unity terms would be the Organization of American States (OAS) because it is the only organisation that includes all countries of its continent and it is the oldest integration organisation in the world, with its origins dating back to 1889 when the I International Conference of American States was held in Washington. Eighteen American countries participated and resolved to create the International Union of American Republics, served by a permanent secretariat called the Commercial Bureau of the American Republics. Later it was to become the "Pan American Union". At the IV International Conference of American States, held in Buenos Aires in 1910, the name of the organisation was changed to OAS.

The OAS process had implications for European integration and Latin American and Caribbean integration. An interesting point to note is that when Latin American and Caribbean countries created CELAC they presented two main arguments, one political and another historical. The political arguments were that the majority of countries did not feel represented in the OAS. From a historical perspective, the justification was fulfilling Bolivar's dream to create a League of American Republics as he proposed in the Congress of Panama in 1826. As we can see, Bolivar proposed a continental league, not a regional or sub-regional union. Bolivar never talks about Latin American countries, always he talks of a Union of American Republics. For this reason, the OAS in its history claims the Congress of Panama as its historical reference to seek continental union, which is why the I International Conference of American States was held.

It is usually said that the promoters of integration processes in Latin America and Caribbean take the European regional integration process as a model, and that is true in most contemporary cases. But it is also true that the OAS model served as an inspiration to the promoters of European integration. The Pan-American movement was a reference point for some fathers of European integration such as Count Richard Coudenhove-Kalergi who wrote his book "Pan Europe" in 1923. In this book, he proposed the creation of a Pan-European movement taking as a reference the Pan-American model (Coudenhove-Kalergi, 2010).

In terms of the history of regional integration, we can observe how there is a parallel history in both regions and how this produces the current EU-LAC inter-regionalism. There is an inter-regionalism of ideas and an inter-regionalism process. The first was developed by thinkers, politicians and diplomats who promoted the ideas of regional integration. The second is an inter-regionalism developed by regional integration organisations.

## 2.4 Science, Technology and Innovation in the EU-LAC Relations

Recently some countries have come to recognise that science, technology and cyberspace have an important role to play in global governance at the same time as it contributes to the strengthening of a country's international presence and its image in the world.

The technological advances produced in areas such as Artificial Intelligence, robotics, nanotechnology, Internet of Things, autonomous vehicles, 3D printing, biotechnology or quantum computing show us that the world has changed. Our societies are experiencing a technological revolution that will modify lifestyle, working and relations (Schwab, 2015). Schwab indicates that the world is experiencing a "Fourth Industrial Revolution" which will produce a convergence between the physical, digital and biological worlds. At the same time, he also has grave concerns that organisations might be unable to adapt, governments could fail to employ and regulate new technologies to capture their benefits and he remarks that shifting power will create important new security concerns, inequality may grow and societies could fragment (Schwab, 2016).

On 18 May 2010 in the framework of the EU-LAC Summit held in Madrid an Action Plan 2010–2012 was approved which defined the following key areas:

1. Science, research, innovation and technology.
2. Sustainable development, environment, climate change, biodiversity and energy.
3. Regional integration and interconnectivity to promote social inclusion and cohesion.
4. Migration.
5. Education and employment to promote social inclusion and cohesion.
6. The world drug problem.

According to the "Science, innovation and technology" research area, the main objective was developing the "EU-LAC Knowledge Area" through:

(a) Improving cooperation in research and innovation.
(b) Strengthening scientific and technological capacities and infrastructures.
(c) To enable sustainable research, innovation and knowledge sharing taking into account the contribution of ancestral and traditional knowledge.
(d) To boost the use of new technologies and technology transfer underpinning sustainable socio-economic development.
(e) To foster cooperation between both regions as regards the digital economy and the reduction of the digital divide for improving competitiveness while making social inclusion a cross-cutting issue.

The Action Plan included a new bi-regional mechanism, the "EU-LAC Joint Initiative on Research and Innovation" later known as "JIRI". The idea was to establish a regular bi-regional dialogue on science, research, technology and innovation and to update common priorities, policies and ensure the proper implementation and effectiveness of cooperation instruments (Council of the European Union 2010). Activities planned included the nomination of "Senior Officials" and alternates in each EU-LAC country by early 2011. The Senior Officials Meetings (SOMs) will be convened annually. These meetings have been the unique bi-regional mechanism institutionalised and permanent from the creation until now. The SOMs were not interrupted despite the fact that EU-CELAC Summits were canceled in recent years. In this case, we can affirm that scientific cooperation does not occur thanks to political dialogue, rather we can say that political dialogue is maintained thanks to scientific cooperation.

In 2013 began the "Network of the European Union, Latin America and the Caribbean Countries on Joint Innovation and Research Activities" (ERANet-LAC) with the aim of supporting the political process of implementing the JIRI. The project period was 3.5 years but was extended to 2017. During this period ERANet-LAC organised three bi-regional joint calls and six Pilot Coordination Actions. The joint calls were realised in close interaction with 35 EU and CELAC funding agencies and in cooperation with other bi-regional initiatives such as ALCUE NET and thematically focused initiatives such as EU-LAC Health, ENSOCIO LAC and NMP-DELA (European Commission, 2022).

Within the framework of the ERA-Net-LAC alone 318 projects were proposed and 55 funded. On 24th January 2022 there was a new call (FECYT, 2022). In 2016 within the SOMs, the Common Research Area (CRA) was adopted and incorporated into the Brussels Declaration. The CRA focuses on three pillars:

1. Increased mobility of researchers between the two regions.
2. Promotion of access to research infrastructures of global interest to improve sharing of data and scientific excellence.
3. Focus cooperation on jointly defined grand challenges to create a critical mass.

In 2017 in the VI SOM four new initiatives and mechanisms were implemented for CRA:

- EU-LAC Working Group on Research Infrastructures (EU-LAC WG RI) to impulse the development of the CRA and to support policymaking in the research infrastructure dimension and facilitate multilateral initiatives (ResInfra EU-LAC, 2022).
- Debate on the development of EU-CELAC mobility portals, building on the existing Euraxess network and portal.
- Agreement on research and innovation actions for bi-regional cooperation in health, climate action and sustainable urbanisation areas (ALCUE NET, 2022).

From a diplomatic perspective, the Global Strategy for the EU Foreign and Security Policy (European External Actions Service, 2016) presented by Federica Mogherini, High Representative of the Union for Foreign Affairs and Security Policy and Vice-President of the European Commission (from 2014 to 2019), makes explicit reference to strengthening EU-CELAC relations and also to science diplomacy (Selleslaghs, 2017). According to Ruffini science diplomacy reduces political tensions and mitigates conflicts between countries, improves mutual understanding and advances the satisfaction of common interests and contributes to global peace (Ruffini, 2020). The academic literature suggests that nations have different motives and expectations of science diplomacy and there are multiple interpretations and conceptualisations of science diplomacy (Knight, 2022), but the praxis show us that it is a valuable foreign policy instrument and for this reason governments (at national or supranational levels) use it in their external action (in bilateral, regional or inter-regional cooperation).

On 30 April 2021, the European Commission approved the Decision C (2021) 2922 which defined the EU-CELAC 20021–2023 Strategic Roadmap for the implementation of the Brussels Declaration and EU-CELAC Action Plan on Science, Technology and Innovation. Both regions reviewed the developments in the EU-CELAC Common Research Area under three pillars: mobility of researchers, cooperation of research infrastructures and global challenges as well as the actions oriented to health, green transition (energy, climate change, biodiversity and food security) and digital transformation with emerging technologies and the digital transformation.

The mobility of researchers at bi-regional levels contributes to creating ecosystems based on knowledge transfer and intercultural understanding between the

EU-CELAC countries. The bi-regional dialogue underlined the importance of human capital development through training and mobility, capacity building of research and innovation organisations from all sectors and creation of doctoral programmes and post-doctoral fellowships within the framework of bi-regional cooperation. A good example is the ResInfra project, its main objective being the construction of a bi-regional collaboration in activities such as a map of National and Regional Research Infrastructures polices to identify partners.

## 2.5 Problems Hindering EU-CELAC Collaboration

As we can observe in the previous section, there are many initiatives and mechanisms for bi-regional level cooperation linked to science, technology and innovation. But there is no initiative or project on cyberspace despite both regions facing the same global challenges of cybersecurity, Internet regulation and the lack of international standards in cyberspace.

There seems to exist a cooperation culture in science diplomacy, in particular at the institutional level, i.e. governmental agencies and ministries. The proof is the collaboration developed between 16 European funding agencies and 15 Latin American and Caribbean funding agencies that participated in the bi-regional calls and although it is true that technology is a transversal area, there is no clear debate on techdiplomacy and cyberdiplomacy.

According to the Forbes Top 100 Digital Companies, among the Top 10 are: Apple, Microsoft, Samsung Electronics, Alphabet, AT&T, Amazon, Verizon Communications, China Mobile, Walt Disney and Facebook (Forbes, 2021). Of these 10 companies, 8 companies are from the USA, 1 from South Korea and 1 from Hong Kong. If we consider the Top 20 digital companies, we can find 12 from the USA, 2 Chinese, 2 Japanese, 1 German and 1 from Taiwan. That is to say, in the TOP 10 digital companies there is not any European, Latin American or Caribbean company. In the Top 20 just we can locate 1 European company from Germany, and we cannot find any Latin American or Caribbean companies. However, the Top 10 or Top 20 digital companies operate in Europe, Latin America and the Caribbean. There is no regulation. In most cases, the goods and services are not included in national legislation and the courts do not have legal instruments to sanction any type of irregularity.

In Latin America and the Caribbean, countries have created specific ministries or strengthened specialised institutions in science, technology and innovation. But there is no budget with investments to support research and development of new technologies. Latin American and Caribbean countries prefer to be importers of technology instead of producing goods and services with high technological content. It is not a lack of natural resources, financial resources or human resources, it is a lack of political will. Although there are government ministries and agencies dedicated to the promotion of science, technology and innovation, the articulation between government–private sector–academia is almost non-existent.

For decades, Latin American and Caribbean politicians have talked about creating value-added products and services, and stopping exporting basic products such as natural resources, which other countries then transform into value-added products, and which later return to the region as imported products. In this sense, the sub-regional and regional integration organisations had to be the ones to defend natural resources production and manage them for the common interest. However, the populist discourse of the defence of national sovereignty is always an argument that ends up not giving up powers to the integration organisations, turning them into mere forums or international meetings without any transcendence. The countries issue declarations of intent, but no political declaration that includes concrete actions.

During 2022 Argentina held the Pro Tempore Presidency of CELAC. Among its main actions under its mandate it planned organising a CELAC-SELA seminar with the theme "Strategies for Post-Pandemic Economic Recovery in Latin America and the Caribbean". The conceptual note of the seminar describes how there will be a session called "Generating spaces for the commercial promotion of MSMEs in Latin America and the Caribbean" whose objectives include (Ministry of Foreign Affairs, International Trade and Worship of Argentina, 2022):

1. Generate a space for discussion to identify strengths and weaknesses in the design of the regional investment and trade promotion strategy.
2. Promote the identification of regional productive complementarities to strengthen the exportable supply.
3. Analyse and discuss good international practices to improve business roundtables in Latin America and the Caribbean.

The Panama Agreement (1975), which constituted the creation of the Latin American Economic System, established in its Article 2 that SELA is a permanent regional organisation for consultation, coordination, cooperation and economic and social promotion, with international legal personality, made up of Latin American sovereign states. This agreement was signed by 26 countries, both Latin American and Caribbean (OAS, 2022).

Article 5 establishes the objectives of SELA to promote regional cooperation in order to achieve comprehensive, self-sustaining and independent development, fostering an improvement in the use of the region's human, natural, technical and financial resources, through the creation and promotion of Latin American multinational companies. It also specifies that said Latin American multinational companies may be constituted with contributions from state, para-state, private or mixed capital. It also promotes the transformation of raw materials from the Member States in the region, seeking industrial complementarity and the export of manufactured products.

The Declaration of the Summit of the Unity of Latin America and the Caribbean (2010) was signed in the Rivera Maya on 22–23 February 2010. The 33 heads of state and government who signed the said declaration (which is not an act or constitutive treaty) agreed on the following objectives:

- Promote regional integration with a view to promoting sustainable development.

- Promote political agreement, boosting the agenda of Latin American and Caribbean in global forums, and a better positioning of Latin America and the Caribbean before relevant international events.
- Promote dialogue processes with other States, groups of countries and regional organisations, to strengthen the presence of the region on the international stage.
- Promote communication, cooperation, articulation, coordination, complementarity and synergy between the sub-regional organisations and institutions.

In the SELA framework, from 1975 to 2023, 58 years have passed. Since the First Declaration of CELAC in 2010 have elapsed 23 years. Latin American and Caribbean countries continue to use the same political discourse but without taking political action. The stagnation of regional integration processes is widespread throughout the region. CELAC does not even have legal personality or competence to execute the agreements that it could negotiate on behalf of the region. SELA does have powers to do so, however, we see how there is a multiplicity of organisations that overlap in objectives and actions, but in none of them do the countries take action. Phrases such as the importance of inserting the region into the international economic system or expanding the region's participation in the main international forums are frequent slogans in presidential speeches, year after year, decade after decade, as we can see.

Among the main problems to be solved in the collaboration between the EU and CELAC on the side of CELAC is the definition of its status and whether it is a regional integration body. The regional integration that the Latin American and Caribbean people are seeking seems not to be the same as the regional integration that the Europeans have developed. It need not be. But the repeated presidential speeches affirming time after time the aim of regional unity and the development of an organisation that can represent the whole region, but without giving the organisation the powers and finance it needs, must stop. Instead, we see how the large and medium-sized economies of the region prefer to negotiate bilateral agreements and achieve their national interests outside the regional organisations they have created. This double discourse of regional integration must be clarified. Otherwise, it will be difficult to carry out the agreements reached with the European Union, which does have an institutional integration system.

In 2021, the Economic Commission for Latin America and the Caribbean (ECLAC) and the Government of Argentina organised the III Meeting of the Conference on Science, Innovation and Information and Communications Technologies. At the meeting, the countries agreed to carry out a series of political and technical meetings, studies and technical assistance activities in the framework of the Biennial Programme of Regional and International Cooperation Activities for 2022–2023 on science, innovation and Information and Communications Technologies ECLAC (2021) based on three strategic objectives:

1. Strengthen public institutions to support the development of science, technology and innovation.
2. Improve the linkage of science, technology and innovation policies with the region's strategic challenges.

3. Foster regional and international cooperation in science, technology and innovation.

In order to advance these strategic aims the countries proposed training programmes in science, technology and innovation policies; exploring national and international experiences and practices that identify critical factors in innovation systems for the design of science, technology and innovation instruments and policies as well as facilitating and promoting spaces for dialogue between government authorities, universities and the productive sector in order to identify concrete actions (Buenos Aires Declaration, 2021).

The pandemic has exposed grave asymmetries between developed and developing countries in the digital economy. The pandemic caused by COVID-19 has produced a new gap, a digital gap. This gap deepens the traditional economic and social gaps. In this sense, the EU-LAC inter-regionalism can serve as a mechanism to design and implement new policies that respond to the times in which we live. The global scope of the pandemic has shown that national responses to the health and socioeconomic consequences of this crisis are not enough (OECD, ECLAC, CAF and European Commission, 2021).

EU-LAC relations have a large number of cooperation programmes on sustainable development, scientific and technical exchange, social cohesion and public security sector. However, it needs to focus on challenges like the green transition or digital revolution (Ayuso, 2021). In this context, EU-CELAC collaboration faces the same global challenges in designing new laws that promote and encourage investments in the tech sector, also called Industry 4.0, and in developing new education plans at all levels to adapt them to a digital world. EU-LAC can promote an inter-regionalism 4.0 for a governance of cyberspace and create a governance 4.0 (Torres Jarrín, 2022).

## References


Ayuso, A. (2021). *Keys to reactivating the EU-CELAC partnership and channeling the globalization of the future*. Barcelona. CIDOB notes internacionals 247. Retrieved December 1, 2021, from file:///C:/Users/Mario%20Torres/Downloads/247_ANNA%20AYUSO_ANG.pdf

Alcue Net. (2022). *EU-CELAC Common Research Area*. Retrieved December 1, 2021, from http://alcuenet.eu/Common_Research_Area.php

Balassa, B. A. (1961). *The theory of economy integration*. Richard D. Irvin.

Bahía Declaration. (2008). *Primera Cumbre de América Latina y el Caribe sobre Integración y Desarrollo*. Salvador de Bahía, 16 y 17 de diciembre de 2008. Caracas. Sistema Económico Latinoamericano. Retrieved December 12, 2021, from http://walk.sela.org/attach/258/EDOCS/SRed/2009/01/T023600003287-0-Primera_Cumbre_CALC_-_Declaraci%C3%B3n_de_Salvador,_Bah%C3%ADa,_Brasil_-_17_de_diciembre_de_2008.pdf

Briceño-Ruíz, J., & Rivarola Puntigliano, A. (2021). *Regionalism in Latin America: Agents, system and resilience*. Routledge.

Buenos Aires Declaration. (2021). *Third meeting of the conference on science, innovation and information and Communications Technologies of the Economic Commission for Latin America and the Caribbean*. ECLAC virtual meeting. Retrieved December 1, 2021, from https://innovalac.cepal.org/3/sites/innovalac3/files/21-00974_ccitic.3_buenos_aires_declaration.pdf

Cancun Declaration. (2010). *Latin American and Caribbean Unity Summit Declaration*. Retrieved December 1, 2021, from https://www.europarl.europa.eu/intcoop/eurolat/key_documents/cancun_declaration_2010_en.pdf

Caracas Declaration. (2011). *Declaración de Caracas "En el bicentenario de la lucha por la independencia hacia el camino de nuestros libertadores"*. Caracas, 2 and 3 december 2011.

Cantori, L. J., & Spiegel, S. L. (1970). *The international politics of regions*. Prentice-Hall.

Cottey, A. (2009). Sub-regional cooperation in Europe: An assessment. *Bruges Regional Integration & Global Governance Papers*, UNU-CRIS and College of Europe Paper 3/2009. Retrieved December 1, 2021, from http://aei.pitt.edu/33535/

Coudenhove-Kalergi, R. N. (2010). *Paneuropa. Raíces de Europa*. Universidad San Pablo CEU.

Declaración de la Cumbre de la Unidad de América Latina y el Caribe. (2010). *Comunidad de Estados Latinoamericanos y Caribeños*. Sistema Económico Latinoamericano y Caribeño. Retrieved September 4, 2022, from www.sela.org

De Lombaerde, P., & Söderbaum, F. (Eds.). (2013). *Regionalism* (Vol. 1–4). Sage.

ECLAC. (1971). *Documento Informativo. Comisión Especial de Coordinación Latinoamericana*. Constitución, resoluciones, declaraciones adoptadas hasta enero de 1971. Santiago de Chile. Repositorio CEPAL (pp. IV).

ECLAC. (2021). *Countries advocated for Science, Innovation and New Technologies to play a more active role in policies for the region's economic, productive, and social development*. Retrieved December 1, 2021, from https://www.cepal.org/en/pressreleases/countries-advocated-science-innovation-and-new-technologies-play-more-active-role

European Council. (2015). "Political declaration". EU-CELAC Summit. Brussels 10–11 June 2015.

European Council. (2016). *Council conclusions on Venezuela*. Press releases. 18 July 2016.

European Commission. (2022). *Final Report Summary-ERANET-LAC (Network of the European Union, Latin America and the Caribbean Countries on Joint Innovation and Research Activities)*. Retrieved December 1, 2021, from https://cordis.europa.eu/project/id/609484/reporting/es

European External Action Service. (2016). *Share vision, common action: A stronger Europe*. A Global Strategy for the European Union's Foreign and Security Policy. Retrieved April 21, 2023, from https://www.eeas.europa.eu/sites/default/files/eugs_review_web_0.pdf

FECYT. (2022). *EU-CELAC platform. Joint actions, statistics*. Retrieved December 12, 2021, from https://www.eucelac-platform.eu/statistics

Forbes. (2021). *Forbes Top 100 Digital Companies*. Retrieved December 1, 2021, from https://www.forbes.com/top-digital-companies/list/

Farrell, L. (2005). Regionalism from a historical perspective. In B. Hettne, L. Van Langenhove, & M. Farrell (Eds.), *Global politics of regionalism. Theory and practice*. Pluto Press.

Fawcett, L. (2005). The origins and development of regional ideas in the Americas. In L. Fawcett (Ed.), *Regionalism and governance in the Americas* (pp. 27–51). Palgrave Macmillan.

Gómez Hernández, J. (2020). Los procesos de integración regional en Europa y América Latina (1951–2007). *Revista de Estudios Europeos*, N. 75 (pp. 350–364). Instituto de Estudios Europeos. Universidad de Valladolid.

Gardini, G. L., & Ayuso, A. (2015). EU-Latin America and Caribbean inter-regional relations: Complexity and change. Barcelona: CIDOB. Retrieved December 1, 2021, from https://www.cidob.org/en/publications/publication_series/project_papers/atlantic_future_papers/scientific_paper/eu_latin_america_and_caribbean_inter_regional_relations_complexity_and_change

Haas, E. B. (1958). *The uniting of Europe: Political, social and economic forces, 1950–1957*. Stanford University Press.

Hänggi, H. (2000). *Interregionalism: Empirical and theoretical perspectives*. Paper prepared for the workshop "dollars, democracy and trade. External influence on economic integration in the Americas". Los Angeles CA.

Hettne, B., Inotai, A., & Sunkel, O. (2002). *Globalism and the new regionalism*. Palgrave Macmillan.

Hettne, B., & Söderbaum, F. (2002). Theorising the rise of regionness. In S. Breslin, C. Hughes, N. Phillips, & B. Rosamond (Eds.), *New regionalisms in the global political economy*. Routledge.

Irela. (1996). *Informe sobre las relaciones entre la Comunidad Europea y América Latina*. Europa-América Latina: 20 años de documentos oficiales (1976–1996). Instituto de Relaciones Europeo-Latinoamericanas. Madrid (p. 7).

Katzenstein, P. J. (2005). *A world of regions: Asia and Europe in the American imperium*. Cornell University Press.

Knight, J. (2022). *Knowledge diplomacy in international relations and higher education* (p. 76). Springer.

Lafta. (1964). *Texto de la Declaración Conjunta de la Misión del Parlamento Europeo y del Comité Ejecutivo Permanente de la Asociación Latinoamericana de Libre Comercio* 16 de marzo de 1964. Documentos informativos 23–78. 4. VI 62–9 VII 64. Archivo ALADI.

OAS. (2022). Convenio Constitutivo del Sistema Latinoamericano. *Convenio de Panamá*. Organización de Estados Americanos. Tratados Multilaterales. Retrieved September 4, 2022, from http://www.oas.org/juridico/spanish/firmas/f-45.html

Ministry of Foreign Affairs, International Trade and Worship of Argentina. (2022). Seminario CELAC-SELA *"Estrategias para la Recuperación Económica post Pandemia de América Latina y el Caribe"* Concept Note. CELAC, SELA, CELAC 2022 Argentina. Retrieved September 4, 2022, from https://www.cancilleria.gob.ar/userfiles/ut/seminario_celac_-_sela_estrategias_para_la_recuperacion_economica_postpandemia_-_concept_note_y_agenda_sesiones_i_y_ii.pdf

Mitrany, D. (1975). *The functional theory of politics*. School of Economic and Political Science.

Mundell, R. (1961). A theory of optimum currency areas. *The American Economic Review, 51*, 657–665.

Nye, J. S. (1971). *Peace in parts: Integration and conflict in regional organizations*. Little Brown & Co.

OECD, ECLAC, CAF and European Commission. (2021). *Perspectivas económicas de América Latina*. Avanzando juntos hacia una mejor recuperación (p. 11). Retrieved December 1, 2021, from https://www.oecd.org/dev/Perspectivas-economicas-America-Latina-2021-Overview-ES.pdf

Puchala, D. (1968). The pattern of contemporary regional integration. *International Studies Quarterly, 12*(1), 38–64.

Rivarola Puntigliano, A., & Briceño-Ruíz, J. (2013). *Resilience of regionalism in Latin America and the Caribbean. Development and Autonomy*. Palgrave Macmillan.

ResInfra EU-LAC. (2022). *EU-LAC cooperation*. Retrieved December 1, 2021, from https://resinfra-eulac.eu/eu-lac-cooperation/

Russett, B. M. (1967). *International relations and the international system: A study in political ecology*. Rand McNally.

Ruffini, P. B. (2020). Introduction to the forum on science diplomacy. *The Hague Journal of Diplomacy, 15*(3), 355–358.

Schmitter, P. (1970). *A revised theory of regional integration, international organization* (pp. 836–868). Cambridge University Press.

Schwab, K. (2015). *The fourth industrial revolution. What it means and how to respond. Foreign Affairs*. Council on Foreign Relations.

Schwab, K. (2016). *The fourth industrial revolution*. World Economic Forum.

Selleslaghs, J. (2017). *EU-Latin America Science Diplomacy*. EL-CSID Working Paper Issue 2017/8, September 2017. Retrieved April 21, 2023, from http://aei.pitt.edu/92500/

Torres Jarrín, M. (2022). Techplomacy. El Interregionalismo de la Unión Europea-América Latina y el Caribe en la Era Digital: Un interregionalismo 4.0 para una gobernanza 4.0. In P. Astroza

& B. Larraín (Eds.), *Relaciones entre la Unión Europea y América Latina: Escenarios futuros en un mundo de cambios* (pp. 69–95). Tirant lo Blanch.

Torres Jarrín, M. (2018). EU-CELAC: A multiplayer interregionalism. Redefining the Atlantic area. In G. L. Gardini, S. Koschut, & A. Falke (Eds.), *Interregionalism and the Americas*. Lexington Book.

Viner, J. (1950). *The customs union issue*. Carnegie Endowment for International Peace.

# Chapter 3
# Science Diplomacy

**Abstract** The chapter will seek to provide a solid conceptual basis for science diplomacy, setting it within the context of international relations and diplomacy studies. It will explore the three axes along which science and diplomacy engage (agency, process and subject matter), drawing on historical and current examples of all three. It will examine the relations between scientists and diplomats and how these are changing in the twenty-first century. It will set science diplomacy in the context of the twenty-first century geopolitical and scientific environments, exploring how it is affected by changes in both, including in the relative international balance of power.

**Keywords** Science · Diplomacy · Science diplomacy · Agency · Process · Geopolitical environment

## 3.1 Introduction

There is nothing new in the engagement of diplomacy with science, or indeed the prominence of science in international relations. International meetings of scientists spanning international borders were common throughout the twentieth century. The successive Solvay conferences in the first half of the twentieth century brought together the leading physicists to discuss the latest developments in their field ranging from relativity to quantum mechanics and atomic physics. The international networks of contacts developed in these conferences played their part in the decisions by the UK and the USA to develop nuclear weapons. The same networks ironically also came into play in the anti-nuclear movement after WWII. This culminated in the Einstein-Russell manifesto of 1955 in which Albert Einstein and Bertrand Russell urged world leaders to think in different ways about nuclear weapons. This manifesto in turn served as the inspiration for Sir Joseph Rotblat to convene the first "Pugwash Conference" in the village of Pugwash in Nova Scotia in 1957. The Pugwash conferences bring together scientists, policymakers and other



M. Torres Jarrín, S. Riordan, *Science Diplomacy, Cyberdiplomacy and Techplomacy in EU-LAC Relations*, United Nations University Series on Regionalism 25, https://doi.org/10.1007/978-3-031-36868-4_3

experts to explore ways of eliminating nuclear and other weapons of mass destruction. Pugwash conferences still meet today and won the Nobel Peace Prize in 1995. This example of scientists actively working together to shape and influence international security policy would appear to be a clear example of science diplomacy in action.

This would be traditionally described as Track 2 diplomacy, activity by and between non-state actors to shape the debate on international issues. But science and scientists have also influenced and played an active role in what would appear to be Track 1 diplomacy—diplomatic negotiations between state actors. Many multilateral negotiations focus on highly technical issues which require the advice of technical experts. This is reflected in the specialised agencies of the UN, for example the Food and Agriculture Organisation (FAO), the International Civil Aviation Organisation (ICAO), the International Fund for Agricultural Development (IFAD), the World Health Organisation (WHO), the World Meteorological Organisation (WMO), the International Telecommunications (ITU) or the International Atomic Energy Agency (IAEA). All these bodies are essentially structured along intergovernmental lines, like the UN itself, and delegations are often led by career diplomats from foreign ministries. But technical experts are essential to the working of these organisations. Scientists and other technical experts will often find themselves members, if not leaders, of national delegations in these organisations, especially in negotiating specific topics (Gluckman et al., 2017).

The creation of the European Centre for Nuclear Research (CERN from its French initials) (Flink & Schreiterer, 2010) represents one of the major European diplomatic achievements in the area of science. CERN has 23 European members and 3 European associate members in the pre-membership stage. It also has eight associate members including non-European countries like India and Pakistan. Other states like the USA enjoy observer status for some of CERN's activities, for example in relation to the Large Hadron Collider (LHC). A broad range of countries enjoy collaboration agreements with CERN allowing cooperation in experiments including in relation to the LHC. The LHC is the most prominent but not the only experimental facility at CERN. CERN demonstrates diplomatic activity in relation to science in at least two respects: the diplomacy required for the construction or CERN and CERN's role in constructing international information exchange and collaboration networks. The negotiations between CERN's founding members on the creation and location of the organisation required considerable diplomatic skill backed up by scientific expertise. Both were required even more in the decision to construct the LHC, a significant financial contribution by all member states. It is striking that while a broad coalition of European states was able to take this decision and see it through successfully, a single government, that of the USA, eventually abandoned its efforts to build a similar capacity. But CERN has also leveraged its scientific success and reputation to build international networks of influence and information exchange. This strengthens CERN itself and, through what looks very much like public diplomacy activities, allows CERN to influence international debates. This is one clear area of scientific activity where Europe, through its ability to work together and support CERN and the LHC despite political contretemps like Brexit, has become and remains a world leader.

The need for technical and scientific advice was not limited only to multilateral organisations. Bilateral negotiations between states often include a strong technical or scientific element. The author of this chapter was involved in the negotiations between the UK and China on the future of Hong Kong following the signing of the 1984 Joint Declaration. Much attention was focused on Hong Kong's continued participation in international organisations, especially technical/scientific organisations like ICAO or the WHO. Such negotiations involved two separate sets of issues. The first related to Chinese sensitivities about sovereignty and the international status of Hong Kong after its return to China. This was the realm of the diplomats. The second was the scientific and technical benefits that would accrue to Kong Kong (and indirectly to China) from Hong Kong's autonomous participation in these organisations. This required the advice and support of the technical and scientific advisors. Often sensitivities of the first kind could be overcome by exchanges between scientists and technical experts. Whether this would amount to "science diplomacy" or scientific support for standard bilateral diplomacy will be discussed later in the chapter.

Scientific knowledge has long informed international negotiations and diplomacy. The pre-WWI conventions which sought to ban certain kinds of weapons (for example dum-dum bullets—hollow-tipped bullets which exploded on impact causing horrendous wounds) were the beginning of a long series of arms reduction talks dependent on technical advice. Indeed, by the later Cold War a group of specialist arms control negotiators emerged who combined geopolitical awareness with diplomatic skills and scientific and technical knowledge. In the same way, negotiations for international treaties like the Antarctic Treaty required a similar combination of diplomacy, geopolitical awareness and technical/scientific knowledge. In the case of the negotiations for the UN Convention on the Law of the Sea (UNCLOS) they were joined by the international lawyers. Scientists have even been able to communicate across the battlefields. In 1941 German nuclear physicist Heisenberg travelled to Copenhagen to talk with Danish physicist Bohr. What was said in the meeting remains controversial, but Heisenberg claims he was trying to get a message to the Allies about the German nuclear weapons programme. Similar attempts at communication by Western atomic scientists at the end of WWII were regarded as espionage. We have already seen how, during the Cold War, the Pugwash Conferences brought scientists together from both sides to promote dialogue and nuclear disarmament. The scientists sought to influence the geopolitical debates. But at the same time, both sides' diplomats sought to manipulate the scientists to influence the narrative. It is not simply a case of scientists advising or seeking to influence policymakers. Scientists can be a tempting tool for diplomats and intelligence officers too in the right circumstances. Multilateral organisations like UNESCO and national cultural organisations like the British Council sought to promote international scientific exchanges and collaboration for their own value in themselves. But debates within UNESCO in the 1980s, especially under the shadow of the withdrawal of the USA, the UK and Singapore, focused on the extent to which these networks and other UNESCO activities were being manipulated by the Soviets as part of their disinformation operations. The USA–Russian collaboration on the International Space Station was symbolic of the end of the Cold War. By the end of the century, scientific and environmental attachés were common in Western embassies (the British Embassy in Beijing acquired an environmental attaché as early as 1989). Such attachés combined career diplomats who had been briefed on the scientific issues in their portfolio with scientists on temporary attachment to embassies. And yet, despite the close engagement of science and diplomacy there was no single area of either practice or study that could be defined as science diplomacy.

This has changed only in the twenty-first century when science diplomacy has emerged as both an activity and an area of study (Flink, 2021). Science diplomacy programmes and activities have been developed at multilateral, national and subnational levels. At the European level, the EU created the S4D4C (Science for/in Diplomacy for addressing global Challenges) consortium in 2018 (S4D4C, 2019). The programme ended in 2021 and has now evolved into the EU Science Diplomacy Alliance. The Madrid Declaration on Science Diplomacy signed by a broad range of science diplomacy activists in February 2021 "aims to foster agreement and raise awareness about the need to strengthen science diplomacy strategies and practices worldwide for the support of universal scientific and democratic values". Cities like Geneva, Barcelona, New York, Shanghai and Mexico have sought to develop themselves as international science and innovation hubs building international networks between them, allowing science diplomacy to merge with city diplomacy (Gual Soler, 2020). For example, Barcelona has created the Science and Technology Diplomacy Hub (SciTech DiploHub), a public–private partnership that aims to position Barcelona as a global leader in science and technology diplomacy as well as serve as a platform for promoting scientific and technological innovation in Barcelona. The SciTech DiploHub offers a series of events and training courses as well as an annual summer school. At the university level, universities are both developing science diplomacy as an increasingly important area of study and developing their own science diplomacy through international networks of scientific collaboration. Thus, the University of Geneva has created the Geneva Science and Diplomacy Anticipator (GESDA), an international network of scientific and diplomatic experts trying to anticipate the main issues and challenges in science diplomacy over the medium to long term. Science Diplomacy has become a prominent issue at the annual Canadian Science Policy Centre (CSPC) conferences.

## 3.2 The International Relations Context

The new focus on science diplomacy resulted from a series of developments in international relations throughout the 1990s and the first decade of the twenty-first century. These included the emergence of the New International Security Agenda through the 1990s, the increase in the breadth and kind of actors, especially non-state actors, participating in international affairs and the development of new digital technologies, especially information communications technologies (ICT), which facilitated the interactions between the new actors. The New International Security Agenda, sometimes referred to as the Global Issues Agenda, provided the framework for the emergence of science diplomacy as both a practice and an area of study. With the fall of the Soviet Union and the end of the Cold War, there was a general feeling that mankind had moved into a new era where old concepts of international security—based on the security of the state, conflicts over territory and power, zones of influence and balance of power—were no longer relevant, at least in developed countries. Fukuyama heralded a world where conflicts over ideology had come to an end and where convergence towards the dominant liberal market economy model would allow greater international collaboration. International security experts increasingly focused on broader issues that threatened not only individual countries but mankind as a whole. There were often referred to as cross-border or transnational issues. This re-thinking on international security issues was given a major impulse by the 9/11 attacks on the Pentagon and the World Trade Centre. It was realised that if international security was defined only in terms of the security of the state, then these attacks could not really be seen as an international security issue as they had represented no threat to the security or stability of the USA. Broadening the definition of international security to threats to the security and economic welfare of the individual within the state allowed international terrorism to be included within the international security agenda but it also allowed a much broader range of issues to be included too (Riordan, 2003). These included poverty, migration, environmental degradation, financial stability, poverty, climate change and pandemic disease. In the context of the turn of the millennium, and the continued belief in the possibility of international collaboration, the development of the New International Security Agenda made sense. Moreover, the more these issues were studied in a security context the more it became clear that these were security issues in the sense of the more traditional security agenda, threatening the stability and security of the state as well as the security and economic welfare of the individual of the state. Poverty and economic crisis were major driving factors behind the Arab Spring in 2011 which destabilised states across North Africa and the Middle East and caused the collapse of several regimes. The economic crisis had in turn been driven by global financial instability which had almost caused the collapse of the international monetary system. Climate change is driving food shortages and poverty and threatens the viability of states in the Pacific and Global South. It is one of the drivers behind mass migration (together with the consequences of the African Spring and the Afghan War) which in turn provoked a major political crisis in Europe in 2015 (and may do again if the food shortages caused, in part, by the Ukrainian War lead to further political instability in North Africa and the Sahel). Post-COVID-19 the impact of pandemic disease on the global economy is already clear. Less clear yet are the security implications. But previous pandemics have had major impacts on the end of WWI (H1N1—the so-called Spanish Influenza) and the stability of states in Sub-Saharan Africa (HIV-AIDS).

The issues of the New International Security Agenda were found to have several factors in common:

- They are all interconnected and interdependent. They can often reinforce each other creating spirals of ever more serious and ever broader consequences (sometimes referred to as positive feedback loops). For example, the poverty and ignorance that drive farmers to overuse land not only lead to environmental degradation in the form of soil erosion but yet more poverty. This vicious cycle is reinforced by increasing the risk of disease and migratory pressures. Scarce farmland forces farmers into ever closer interaction with wild animal species increasing the risk of zoonotic disease transfer. Migrations both spread disease and increase poverty and environmental pressures in neighbouring communities. The pressures of migration, poverty, environmental degradation and disease call into question the governmental and other institutional structures of a region or a country opening up opportunities for organised crime and international terrorist groups, quite aside from the increased risk of civil or other conflicts. The key point is that none of these issues can be dealt with individually. They need to be tackled as a whole. A holistic approach is needed.
- There are no ready-made solutions or "right answers". Defining the problems precisely can be difficult let alone specifying detailed policy objectives. Expert opinion can disagree radically, with even the broad road to follow in dispute. For this reason, they are often referred to as "sticky" or "wicked" problems. This also makes them like the broader problems of foreign policy, where management rather than solution, is order of the day, and where the best outcome is often the least bad of pretty awful alternatives. Solutions (if they exist) or responses must be developed through dialogue and conversation.
- They are all global issues. No country, or even regional grouping of countries, can solve them on their own. Even if the entire European Union were to achieve negative carbon emissions it would have no impact on climate change if the USA and China continued to pump carbon into the atmosphere. The recent COVID-19 pandemic has demonstrated the weakness of trying to deal with pandemic disease at a national level. International collaboration is essential which means that responses to these global issues cannot be limited by ideology or value-driven agendas. Global collaboration means finding grounds for cooperation even with those you disagree with or dislike.
- Collaboration just between governments is not enough. International collaboration has to extend beyond governments to corporations, NGOs, universities, health and other professionals and civil society actors. COVID-19 has demonstrated the importance of major pharmaceutical companies in tackling pandemic diseases. But it also demonstrated the reluctance of governments to react quickly or communicate effectively. It was not just the Chinese government that was reluctant to communicate what was happening in Wuhan: in March 2020, the Spanish government sought to downgrade the warnings emerging at a European level to save the Easter holiday season. In dealing with many of these global issues direct relations between professionals, whether individuals or within NGOs or other bodies, need to supplement relations between governments. If you want early identification of the next pandemic disease, you may be better off talking to local health officials or paramedics than the Health Ministry. In effect,

inter-government collaboration had to be complemented by inter-civil society collaboration.

- All these issues have heavy scientific or other expert knowledge content. Generalist diplomats cannot handle them without briefings from scientists and other experts. In some cases, briefings will be insufficient, encouraging the incorporation of scientists and other experts into the management of the agenda. But (and this is a theme that will be picked up again in Chap. 6 Science Diplomacy: Lessons from COVID-19 and Climate Change) this is not simply a case of putting scientists in charge of the negotiations. Because the issues of the New International Security Agenda are so interconnected and interdependent, none have straightforward scientific or technical solutions. It is this interconnectedness that makes them "sticky" or "wicked" problems. Managing them involves international politics and messy compromises as well as rational discussion of the scientific basis. Diplomats as well as scientists are needed (Cooper et al., 2008).

A corollary to the emergence of the New International Security Agenda was the re-emergence of more traditional geopolitical agendas. As suggested above, following the collapse of the Soviet Union and the end of the Cold War there was a feeling that geopolitics, if not history itself, had come to an end and that mankind could now focus on the big global issues of the New International Security Agenda. This was particularly strong in Europe, where foreign ministries and diplomatic services were largely reconfigured to tackle these big global issues and where Cooper held up the European Union as the new Post-Modern State of the future. Thus, the Russian invasion of Georgia in 2008 followed by the Russian invasions of Ukraine in 2014 and 2022, not to mention China's militarization of the South China Sea, came as something of a shock. Non-Western countries would make similar points about NATO action in Kosovo or US-led regime change in Iraq and Libya. The upshot is that geopolitics has gone nowhere and conflicts over territory and resources, balance of power and zones of influence are back on the international agenda. This in effect creates a conflict between the two agendas. The extent to which countries have to focus on the more traditional geopolitical agendas undermines their ability to deal with the global issues agenda. Geopolitical issues are short term and urgent—a foreign minister who gets them wrong will suffer the political consequences—while the global issues agenda is more long term—if a foreign minister gets climate change wrong it is his grandchildren who will suffer the consequences. As Chap. 6 explores, geopolitical tensions and conflicts hampered global responses to the COVID-19 pandemic. Recently, China has suspended collaboration with the USA over climate change because of growing geopolitical tensions on Taiwan. The geopolitical agenda as much as the New International Security Agenda forms the background to twenty-first century science diplomacy.

Non-state actors have always been involved in international relations. The myth of a golden age of international relations dominated by states in which diplomats enjoyed a monopoly over foreign policymaking is just that: a myth. Big corporations have always been major international players. The British and Dutch East

India Companies had their own armies and navies. The British East India Company governed India until 1857 when the British Government finally incorporated it into the empire following the Indian mutiny. The major oil companies in the twentieth century in effect had their own foreign ministries and intelligence services within their corporate structures. The first NGO was probably the Anti-Slavery League (founded in the eighteenth century and still going strong). NGOs like Amnesty International and International PEN were active throughout the Cold War. But the development of digital technologies, and in particular the Internet, and the emergence of the New International Security Agenda, have led to an explosion of non-governmental actors able to participate in international debates in the first decade of the twenty-first century.

Digital technologies have facilitated three capabilities which non-state actors did not enjoy, or only in limited ways, before: access to new sources of information and analytical techniques; the ability to network across national borders with like-minded people and organisations; and the ability to campaign at a global level. The World Wide Web, online media sources and search engines allow non-state actors to access information that previously would have been limited to diplomats and other state actors. Satellite images that were once highly classified are now available through commercial satellite companies often for minimal cost. Drone technologies further enhance the scope of non-state surveillance. Relatively cheap drones can live stream video from areas that would otherwise be inaccessible to non-state actors because of physical or state security. Such technologies are already being extensively used by human rights and environmental NGOs. This range of techniques and technologies is increasingly being brought together in what is known as Open-Source Intelligence (OSINT). OSINT combines technologies like drones with the vast quantities of information freely available ("open source") on the internet to monitor the behaviour of state actors and their agencies. Organisations of independent journalists like Bellingcat are able to use these techniques to identify criminal behaviour and indict state actors (Riordan & Torres Jarrín, 2020).

In the past, political effectiveness depended on reaching a critical mass of popular support in a limited geographical area. This is no longer the case. Small, scattered groups of activists or enthusiasts can network at regional or even global level achieving levels of influence that would have been impossible at merely local or national level. These networking opportunities have been further enhanced by the development of social media platforms. The same technologies that have supported networking among these non-state groups also facilitate campaigning and the spread of their messages. One consequence has been the development of global non-state activists focusing on a single or a limited range of issues. Traditional political movements prior to digital technologies had to adopt policy positions on a whole range of political issues. In doing so they risked turning off potential supporters who shared some but not all of their policy positions. Now activists can cherry pick the international single-issue organisations that best meet their individual mix of positions. Many of these single or limited-range organisations focus on global justice and inequality and environmental issues. This naturally draws them towards the issues of the New International Security Agenda with their technical and scientific content.

The growth of non-state participation in the debates of the New International Security Agenda strengthens the need for science diplomacy in the management of those issues. This growth of scientific content in non-state (sometimes called Track 2) diplomacy is reinforced by the growth of exchanges between universities and other academic institutions. Confronted by the growing influence of these non-state actors, especially in areas related to climate change and environmental degradation, companies also feel compelled to engage in the debates. Science diplomacy becomes an area in which a broad range of state and non-state actors engage with each other.

## 3.3 Science Diplomacy Along the Three Axes of Diplomacy

The sudden increase in interest in science diplomacy can thus be seen to arise in part from the New International Security Agenda, with its science heavy themes, the multiplication of state and non-state actors and the emergence of new digital technologies all coinciding in the first decade of the twenty-first century. Despite the rapid growth of science diplomacy as both an activity and a topic of study, there remains considerable confusion about what it is. Does it refer to the international activities of scientists or the involvement of diplomats in scientific issues at an international level? Sometimes it seems to refer to little more than efforts by countries or cities trying to brand themselves as centres of science, technology and innovation (STI) in the hope of attracting foreign direct investment (FDI).

In 2010, the Royal Society sought to bring some clarifications and rigour to the confusion by suggesting three kinds of science diplomacy: science in diplomacy, diplomacy for science and science for diplomacy. "Science in diplomacy" refers to the role of science and scientists advising diplomats, or even taking part in diplomacy, on international issues with a heavy scientific content: for example, the role of scientists as advisors or participants in climate change negotiations or the supporting roles we have seen for scientific or technical advisors when negotiating international treaties like the Antarctic Treaty or the UN Convention on the Law of the Sea. This advisory role is likely to grow as diplomacy increasingly has to engage with scientific and technical issues like cyberspace, artificial intelligence, biogenetics and even the "metaverse". "Diplomacy for science" refers to the activities of diplomacy and diplomats in promoting international scientific exchanges or creating an international environment in which scientists can collaborate for example negotiations in the EU about where to build a new scientific research facility or about creating new research funding. It could be argued that such discussions within the EU would be better described as "Bureaucracy for Science" since much of it amounts to internal horse-trading rather than diplomacy as such. However, in the EU-CELA context there is a much clearer argument for a role for diplomats in promoting transatlantic scientific cooperation and even joint EU-CELA scientific research projects. "Science for Diplomacy" refers to the use of science to support broader diplomatic objectives. This might include the use of a reputation for scientific excellence in a public diplomacy campaign to secure foreign direct investment or the location of a major research facility on a country's territory. It could be argued that CERN enhances Europe's reputation as a centre of excellence for experiments in particle physics through the operation of the LHC. CERN's high levels of activity in promulgating its achievements and international networking have undoubtedly raised its own profile and reputation (CERN has even made it into a Dan Brown novel—"Angels & Demons" and the associated film with Tom Hanks) (Flink, 2021).

An alternative approach would be to think of science diplomacy for gaining access to other resources abroad (knowledge, finances or talent), for national promotion (promoting science, technology and innovation capacity (STI) and higher education institutions (HE)) or to exert influence on other actors through the use of STI and HE. The categories are related but emphasis more the instrumentality of science diplomacy. Science diplomacy is undertaken not in the hope of promoting greater international collaboration or addressing global issues (although that might happen coincidentally) but rather in pursuit of narrow national interest. This may indeed be how many political leaders think of science diplomacy (or be how foreign ministries justify it to sceptical finance ministries). But, as we have seen, state actors are not the only participants in science diplomacy. The broad range of non-state actors, from environmental activists to CERN, would protest strongly at having their activities characterised as the pursuit of national or other selfish interests. While CERN, for example would not object to any benefits that European countries or companies might accrue as a bi-product of its activities, it would resist denoting those benefits as its prime objectives. This characterisation of science diplomacy in terms of national interest does, however, draw attention to tensions that run throughout science diplomacy between a multilateral and multistakeholder engagement to promote more equitable international scientific collaboration or using science to promote better international understanding and using science diplomacy to gain material benefits and influence at national (or regional or city) level. These tensions can be seen, for example, in Barcelona's SciTech DiploHub. While the diplomatic entrepreneurs behind this initiative think primarily in terms of international networking and promoting debate on science diplomacy, the Catalan regional government, which provides financial support, is focused more on re-establishing Barcelona's international reputation following the fiasco of the Catalan independence referendum (Gual Soler, 2020).

The problem with the Royal Society's categorisation of science diplomacy is that it focuses on science and pays little attention to diplomacy (perhaps inevitably given the nature of the Royal Society). In particular, it pays little attention to how diplomacy functions in broader contexts. An alternative way of thinking about science diplomacy is to start with diplomacy rather than science. Diplomacy engages with other areas along three axes: agency, process and domain or subject matter. Roughly these three axes can be characterised as who are the diplomats (agency), what skills and techniques are used by the diplomats (process) and to what area or areas is the diplomacy to be applied (domain or subject matter). So in relation to another growing area of diplomacy studies, sports diplomacy, agency would consider whether sportsmen could function as diplomats (e.g. various UNICEF ambassadors, and what that meant in terms of diplomacy), process would explore how diplomats were able to use sport and sports events to promote broader diplomatic objectives (e.g. China's ping pong diplomacy or the use of the Winter Olympics by North and South Korea to promote dialogue) and domain would look at the application of diplomacy to issues arising in sport (e.g. negotiations to calm tensions following the bodyline test series in Australia or campaigns to host major sporting events). Likewise in relation to science diplomacy, agency explores the question of who is doing the diplomacy (in particular, the role of scientists as diplomats), process looks at how diplomats might use science to achieve broader foreign policy objectives and domain examines how diplomacy can be applied to policy problems generated by science or developments with heavy scientific content.

**Agency**

Scientists acting as diplomats is problematic. The Madrid Declaration affirms that Science Diplomacy "goes beyond international science collaboration, as it tackles interests that go beyond the scientific ones and may directly or indirectly serve to advance diplomatic goals". But it is not clear what this does, or could, mean. Science and diplomacy are very different activities, and scientists and diplomats are very different kinds of creatures. Science is based on the idea that there is a single and knowable reality, which scientists can come to know, in theory at least, through a combination of theory and experiment. While it is a caricature to suggest that scientists deal with certainties and diplomats with uncertainty (not least because of developments in quantum mechanics) there is the sense in which scientists are gathering information to test hypotheses. These may not necessarily be in the form of replicable experiments. Some sciences such as geology and evolutionary biology may find replicable experiments impossible, at least in some areas of their work. But there is always the sense that scientists are collecting data to strengthen their convictions in their interpretations of the world. Diplomats do not have that fortune. As Henry Kissinger once said, the tragedy of foreign policy is that when you have the greatest freedom of manoeuvre you have the least information. By the time you have sufficient information you have lost your freedom of manoeuvre. Your options have been closed off (Gewen, 2020). Diplomats have to deal with uncertainty and imperfect knowledge. It is not just that diplomats are unable to reveal the perfect knowledge on which their certain decisions could be built. The very nature of the issues that diplomats must deal with preclude perfect knowledge and certainty at the time of taking decisions. It is only with the advantage of historical hindsight that certainty and perfect knowledge are available to the diplomat, and not always then. This in part reflects the complexity of international relations and the concealed intentions of other actors and in part the difference between science and policy. Science as an activity seeks to discover the world and how it functions as it is. Policy aims to change the world (or on occasion prevent the world from changing).

To put it another way, scientific discourse and diplomatic discourse are radically different. Scientific discourse is based on the verifiable exchange of facts and evidence from which it is intended to derive truth or at least a better understanding of the world. Diplomatic discourse seeks to manage complex problems where key information is unavailable, the intentions of other actors known at best partially and there is no single good outcome. Another tragedy of foreign policy is that there are no good or perfect outcomes: diplomats must hope to find the best among a series of bad outcomes. If ever there was doubt about this the 2022 Russian invasion of Ukraine makes the point: there are no clear good outcomes but rather a series of more or less bad outcomes. The art of statecraft is to identify and pursue the least bad of the options. Climate change helps illustrate the difference between science and foreign policy.

The scientific questions are whether climate change exists and whether it is the product of human activity. The foreign policy questions are what can be done about it at an international level. Diplomacy lies in the engagement with other international actors, seeking to understand their positions and interests and seeking to convince them to pursue the policies we favour. It also involves explaining to government ministers when their favoured outcomes are unrealistic and where possible compromises can be found. Although scientists and activists may despair of the inadequate outcomes the diplomats produce, the diplomats recognise that they are confined to seeking the best of the bad outcomes. Policy questions, unlike scientific questions, do not have even theoretically correct answers. There is rarely if ever only one factor involved, and different actors will put emphasis on different aspects promoting different policies. Whereas scientific hypotheses can, at least in principle, be falsified by subsequent experimentation or data, this is not the case with policy decisions. Even with the benefit of hindsight, and with knowledge of the outcomes, it can be difficult to evaluate policy decisions. This is true even in cases where the failure of policy seems clear cut (and seemed clear cut to some people, at least, at the time). Chamberlain's policy of appeasement was widely criticised by some political figures at the time (e.g. Churchill or Nicolson) and the Munich Agreement in September 1938 which conceded the Sudentenland to Germany condemned by many within a year of its signature. And yet historians continue to debate what real policy alternatives, if any, Chamberlain had the time and whether he had the information and knowledge which would have made alternative policy decisions viable (French, 2022). Science can, and should inform, policy debates, but cannot decide them. We can all agree that climate change is real and urgent, and largely generated by human activity, but there are a large number of possible policy responses, which will not be decided by science alone.

This to some extent clarifies the role of scientists as agents in diplomacy. Scientists can act as diplomats. Some eminent diplomats have had a scientific background. But when scientists engage in diplomacy they do so as diplomats, not scientists. The qualities which define good scientists, hamper their activities as diplomats. This should not be surprising. Those scientists most effective in university faculty politics are successful because they stop being scientists and act as politicians. This is an important distinction. The same person can act in different capacities: as a scientist, as a policy advisor or as a diplomat. A scientist advocating for a given policy is no longer acting as a scientist but as a policy advocate. His background as a scientist gives him credibility but does not guarantee that the policy, he advocates is necessarily better than that recommended by someone with a background in economics or foreign policy. As we shall see in Chap. 6, this is a distinction that was somewhat lost sight of during the COVID-19 pandemic. There is now growing criticism of the British government's scientific advisors that they exceeded their role as advisors to pressure the government into certain policies (primarily lockdowns) without allowing ministers to raise non-COVID 19-related factors like the economic and psychological damage these policies might imply or the risks to people suffering other life-threatening diseases. The scientific advisors are responding by accusing the politicians of abdicating their responsibilities with their claims to be "following the science" and now using that as a way of blaming the scientific advisors for their own failings. Whatever the truth of the matter, the roles of scientists, policy advisors and policymakers seem to have become confused in potentially dangerous ways (Riordan & Torres Jarrín, 2020).

This does not mean that scientists acting as scientists do not have a role in diplomacy or foreign policy. We cannot expect diplomats to be experts in the scientific issues they may have to discuss at the international level (although we can expect a reasonable understanding of the key issues). The science itself is not the responsibility of the diplomat. The diplomat's job is to discuss, negotiate and manage the international policy implications of the science according to decisions taken by her government (or international organisation). A better understanding of the science of climate change is essential to diplomacy seeking to negotiate an international agreement on climate change policy but we cannot expect the diplomat to develop it. This does not mean that leaving the climate change debate to scientists would result in better outcomes. It is not the science that is being debated but the policy responses to the science and, whether climate scientists like it or not, these involve many factors other than the science of climate change. Diplomats find foreign policy difficult not because they have insufficiently scientific brains, or lack broader intelligence, but because foreign policy is difficult. Foreign policy must be formulated and implemented under conditions of uncertainty and imperfect knowledge (and considerable complexity). But that is only the first stage. Implementing foreign policy means engaging with a broad range of state and non-state actors and seeking their agreement. Avoiding conflict in such circumstances almost always involves messy compromises (as demonstrated by the various meetings and agreements on climate change). Imposing certainty on such conditions under the illusion of perfect knowledge is a recipe for disaster.

**Process**

Science Diplomacy as process is less problematic. Diplomats have long used science, and scientists, to pursue broader diplomatic strategies. The nature of science, as an international undertaking dedicated to advancing human knowledge, in itself breaks down national barriers. During the Cold War, both Soviet and Western diplomats took advantage of meetings of scientists to gather information, reduce tensions and spread propaganda. Scientific achievements form an important part of national reputation, and as such are an important soft power resource to be deployed in public diplomacy strategies. Scientific exchanges can also serve to create environments in which diplomatic negotiations can flourish. In the post-modern world of fake news and echo chambers reinforced by social media, science and its reputation for veracity and rigour offers at least the potential for helping to tackle disinformation campaigns.

The international network, or community, of scientists is in itself a global forum for the exchange of views and ideas in a way that can reduce or contain national competition. Climate change offers an interesting example. Following WWII, the British Council, among other cultural bodies, promoted international networks of scientists to promote exchange of scientific knowledge and best practice. As noted above, these networks crossed the barriers of the Iron Curtain. The aims of the network building were not entirely altruistic. National organisations like the British Council hoped also to promote the UK as a centre of scientific research and innovation while using the networks also to promote Western values like freedom of expression and human rights. Whatever the motivation for their construction, and somewhat surprisingly given the confrontations of the Cold War, the networks flourished as ways in which scientists could exchange information and discoveries with international colleagues. These international networks continued to flourish after the Cold War. These were the networks through which growing concerns about climate change began to circulate.

Climate change was being discussed on these networks long before environmental activists and NGOs caught onto the problem. The fact that they were networks of scientists, created for other purposes, enhanced their credibility with both governments and publics. Diplomats were able to take advantage of this when negotiating international political responses. It should be pointed out that there are some similarities between these networks within the international community of scientists and those within the international community of diplomats, although these international communities operate in different ways and towards different ends.

However, the interaction of science with diplomacy as process is not entirely without problems. Science and scientists become instruments for diplomats in the pursuit of broader diplomatic strategies and national interests. While scientific achievement as a soft power asset which public diplomats use to enhance the national reputation is relatively innocent, the calculated use of scientists and their activities to promote broader foreign policy objectives, of which the scientists may not be aware, is more controversial. Although diplomats on both sides of the Cold War divide sought to take advantage of scientific exchanges to promote their broader geopolitical agendas, it is clear that some scientists resented being used in this way, while others were manipulated in ways they were not aware of. Unsurprisingly this caused problems in the relations between diplomats and scientists, particularly when the extent of manipulation, by both sides, emerged at the end of the Cold War.

It is likely that many scientists will still resent being manipulated if diplomats and diplomacy take them for granted, using them to promote foreign policy interests alien to their scientific objectives. This is not just true of scientists. The main objective of Public Diplomacy is to influence foreign public opinion in the belief that foreign public opinion can influence the decisions taken by its government. In practice, this is restricted to key target audiences in foreign publics who are likely to have most influence over the relevant decision makers. One of the main tenets of modern Public Diplomacy is that surrogates are more effective than diplomats in influencing foreign public opinion (after all, as the old saw has it, an ambassador is an honourable man sent abroad to lie for his country). Thus, public diplomats are converted into diplomatic entrepreneurs, matchmaking between key figures in academia, business, the media and NGOs and civil society, bringing them together in important policy discussions, whether through conferences, workshops or, more recently, online platforms. But Public Diplomats using civil society surrogates in this way must beware of the temptation to manipulate them for their own diplomatic purposes.

One of the challenges of Public Diplomacy is to accept that the people a diplomat brings together to influence foreign public opinion have their own views and agendas which must be respected. The same is true of the use of science and scientists to promote broader diplomatic objectives. The diplomats must be open with the scientists and respect their agendas if they do not want to provoke a reaction in which scientists decline to collaborate with diplomats, to a mutual disadvantage (Sharp, 2009).

**Domain or Subject Matter**

Science as a subject matter involves the application of diplomacy to science. This does not mean that diplomacy seeks to resolve scientific problems. Diplomacy is no more relevant to the content of science than science is to the content of diplomacy. Rather diplomacy is applied to the international political and geopolitical problems generated by science and its application through technology. Such problems are not themselves scientific or technological but related to international politics and foreign policy. They do not have scientific or technical solutions and so cannot be left to scientists and technicians to resolve.

This raises the question of whether diplomats (and politicians) have any role in discussing international norms or constraints in pure scientific research, rather than its technological applications. The problem can be illustrated by atomic physics. Unsurprisingly diplomats were not involved in the development of the theory of atomic physics in the 1920s and 1930s, and in particular whether it would be possible to split the atom or create chain reactions. These were purely scientific issues, but issues with real-world consequences. Even at this stage in the 1930s, some physicists were speculating over the possibility of building an atomic bomb. These issues were no longer solely scientific or technical. They had serious implications for statecraft and foreign policy. And yet diplomats and foreign ministries remained largely uninvolved, even after the UK and the USA had taken the decision to develop nuclear weapons. Once the bombs had exploded over Hiroshima and Nagasaki in 1945 the issue of nuclear weapons was no longer primarily scientific or even military. Statecraft and its handmaiden diplomacy entered the fray. The Truman Administration seriously debated placing all nuclear weapons under the control of the IAEA. Once the Soviet Union had also acquired the atomic bomb in 1949, diplomats became as important as the military in ensuring that nuclear weapons were not used again. The nuclear weapons radically changed the context in which the diplomats had to operate while calling on their skills to reduce tensions and, ultimately, the number and types of weapons.

The question here is whether diplomats and policymakers should have been involved at an earlier stage and negotiating constraints on the applications and uses of atomic physics. This question extends beyond the scope of this chapter but raises the question of whether international rules and norms should be negotiated restricting the uses and applications of pure scientific research. Many scientists would object to restrictions on pure scientific research and argue that the scientists themselves are best placed to decide such issues, rather than governments or the messy compromises of diplomats. Nevertheless, there may be areas of scientific research where publics or policymakers decide that some restrictions at international level are required. One example is cloning human beings, where public opinion in Western countries is strongly opposed. In the aftermath of COVID-19, there may be pressure for international regulation of biological research centres and practices like "gain of function" research. Any future agreements on these areas would need to be negotiated by diplomats in multilateral fora identifying the perceived interests of other countries and building on the synergies between the perceived interests of different countries. Scientists will have to inform the discussions and negotiations, but it will be diplomats who will have to reach the agreements.

Science also reveals dangers and problems at a global level which require the action of diplomats and diplomacy. Many of these issues feature on the New International Security Agenda, including climate change, environmental degradation and pandemic disease. As argued above, these are as much security issues as scientific ones. While science is required to understand the issues and suggest policy measures to mitigate their impact, the responses are not purely technical or scientific. In responding to the threat of pandemic diseases governments (and major corporations) have to take into account, and balance, a broad range of different factors. The balance will vary from country to country and region to region. The task of negotiating global agreements requires flexibility, the ability to network effectively with state and non-state actors globally, see the problems through the eyes of the different actors and build coalitions of the willing around shared preferred outcomes. In short, diplomacy—but a diplomacy that limits itself not just to state actors, but is capable of engaging effectively with a broad range of non-state actors. This is increasingly referred to as multi-stakeholder diplomacy. Chapter 7 will look at this in more detail in the context of COVID-19.

Diplomats came late to the day to nuclear weapon research. The same could be argued about digital technologies and artificial intelligence. To a large extent these areas have been seen as technical, and thus left to technicians (and intelligence services). But as Clemenceau said, war is too important to leave to soldiers. Cyberspace and artificial intelligence are too important to leave to technicians.

The essential problems relating to Internet governance and cybersecurity are political and geopolitical. They relate to the state and non-state actors which seek to regulate the behaviour of other actors in cyberspace. They relate to the intentions and motivations of state and non-state actors using Internet tools to coerce, damage or steal. The key challenges of cybersecurity are not building technical defences (although these are important), but understanding what certain actors are seeking to achieve, determining whether these actions are acceptable, and if not persuading them to change their behaviour. Building norms in cyberspace, for example limiting the range of acceptable targets, requires identifying the intentions and perceived interests of other state and non-state actors, developing strategies to win round reluctant actors and identifying shared preferred outcomes on which coalitions can be built are all essentially diplomatic tasks. The same applies to other areas where the development of pure science into technology can threaten international peace and stability and/or economic welfare, like artificial intelligence, biogenetics, quantum computing or nanotechnologies. For example, artificial intelligence research will undoubtedly result in the development of Lethal Autonomous Weapons Systems (LAWS), whatever the arguments against them among the general public or even the scientific community. The role of diplomats will be to negotiate on the base of empathetic identification of intentions and shared preferred outcomes, limitations on their use in conflict. These areas will be explored in more detail in Chaps. 4 and 5.

Finally, as we have seen, the New International Security Agenda risks being undermined by the resurgence of more traditional Geopolitical Agendas. We will look in more detail at the specific case of international exchanges and collaboration over COVID-19 later. The war in Ukraine has already resulted in the cancellation of US-Russian collaboration over the space station. China has responded to increasing tension with the USA over Taiwan by suspending a raft of scientific collaboration including on climate change. This points to another, and possibly the most important, role for science diplomacy: creating geopolitically neutral spaces where scientists can meet to discuss scientific issues and problems. The science diplomats may have little to say on the science in question but can play an important role in allowing the scientists to meet among themselves at an international level without outside political or geopolitical interference.

## 3.4 Science Diplomacy in the Twenty-First Century Geopolitical Environment

We described earlier the development and recognition of the New International Security Agenda in the last decade of the twentieth Century and the first decade of this Century. A certain Western triumphalism emerged at the time encapsulated in Fukuyama's "End of History". Although Fukuyama's Hegelian arguments were more subtle than his detractors have allowed, there was a sense that the ideological debate was over and that Western liberal market economies had triumphed. The triumph of the Western model of the state was accompanied by the triumph of Western values. References increased to the rule-based international order, to which all countries were expected to submit regardless of their own historical or cultural traditions. Few were willing to acknowledge that the acceptance of Western-derived value systems and institutions owed more to the extent of US hegemony over the period 1991–2003 than a genuine convergence between the different branches of humanity. Sociologists sought to use data to demonstrate human progress and in particular how violence in human societies was diminishing. Economists and politicians argued that globalisation was an inevitable process that would inevitably lead to ever greater economic and social convergence and integration. In many ways, this period resembled the last great period of globalisation before WWI when Norman Angell argued that the degree of economic integration between the great powers meant warfare was no longer viable.

This period of optimism and relatively peaceful relations between the major powers (outside Africa at least) allowed policymakers to focus on the global issues of the New International Security Agenda (NISA). This tendency was reinforced by 9/11 after which the main security threat to Western societies was seen as terrorism (itself one of the issues on the NISA) rather than inter-state rivalry. The levels of inter-state collaboration on counterterrorism further encouraged this focus on the NISA, with non-collaborating countries seen as rogue rather than rival states. Western foreign ministries and militaries were reconfigured to the requirements of the NISA, with many traditional diplomatic virtues (geopolitical reporting, diplomatic empathy, interests-based approach to foreign policy etc.) being discarded or downgraded as unnecessary in a world in which all states of goodwill (otherwise referred to as members of the international community) would naturally collaborate in tackling the NISA. Even NATO's Strategic Concept was re-drafted in 2010 to accommodate the demands of the NISA (it was re-drafted again in 2022 to take account of the return of geopolitics and inter-state conflict). Europe's would be foreign service, the External Action Service (EEAS) was from the beginning designed on the premise of the European Union as a post-modern state showing the path to the future. In this optimistic world where state borders were thought to be ever less important, science diplomacy was relatively straightforward. The key issues governments wanted to address had high scientific and technical content. Digital technologies had facilitated the participation of the full range of non-state actors (including universities and scientific research groups) in international debates. Direct exchanges between scientists and technical experts on the issues of the NISA across borders and unaffected by foreign policy considerations were uncontroversial.

In an act of unintentional irony, Norman Angell entitled his 1910 book arguing that economic integration made inter-state warfare futile "The Great Illusion". The optimism of a post-Cold War new world order now looks equally illusory. As noted above, Russia's invasion of Georgia in 2008 and Ukraine in 2014 and 2022 have marked the return of conflict over territory and zones of influence to the Eurasian land mass. As we have seen, China's increasingly overt assertion of sovereignty in the South China Sea and growing tensions between Beijing and Washington over trade, technology and Taiwan undermined the prospects of a globally coordinated response to COVID-19. Brexit and the UK's withdrawal from the European Union has undermined cooperation on scientific research within Europe. As governments and regional organisations focus more on short-term geopolitical agendas, the issues of the NISA slip down the political agenda. The European Commission has announced that it wants to be more geopolitical, although it is unclear what this means. As the energy and cost of living crises in Europe, driven in part by the war in Ukraine, increase in political and social importance calls increase, including within governments, to abandon green energy conversion policies in favour of short-term measures to save European economies from collapse. Europe returns to coal consumption to generate electricity.

This means a far more challenging environment for science diplomacy (Flink, 2021). The idea of scientific exchanges at international level uncontaminated by politics or geopolitics is even harder to achieve. It is not simply that the older geopolitical agendas undermine and downgrade the priority of the NISA, as we have already discussed. Interstate rivalry and competition spread from the military to other areas such as trade, finance and science and technology. Indeed, the dangers of escalation in military conflict between nuclear states make these other areas even more attractive for fighting out geopolitical rivalries. Leadership in science and technology becomes a measure of comparative geopolitical power and success. This nationalism in science and technology, rather than encouraging international collaboration, incentivises competition and attempts not only to maximise your own achievements but also to minimise or undermine those of your rivals. International collaborations are already being cancelled whether in space or in the fight against climate change. Technology transfers are being prohibited in efforts to maintain technological superiority and reduce dependence on supply chains (as we will see in the chapter on Techplomacy). A key challenge for science diplomacy in the twenty-first century will be to maintain the very idea of international rather than regional or national scientific collaboration, creating neutral spaces free from geopolitical pressures where scientists can meet and work together. Europe will have an important role to play in this, CERN being a model both of international scientific collaboration and the networks of influence and information that such an organisation can create. Greater and more institutionalised scientific collaboration between the EU and CELA countries offers the opportunity to demonstrate the benefits of inter- and bi-regional science diplomacy as opposed to the growing scientific nationalism. This would suggest that both the EU and CELA countries need to develop a more effective science diplomacy in relation to each other (Gual Soler, 2021).

## 3.5 Conclusion

The underlying theme of this chapter, and indeed of much of this book, is to lay a greater emphasis on the diplomacy aspect when talking about science diplomacy (or cyberdiplomacy or techplomacy). Much of the existing literature on science diplomacy focuses on science. A subtext of this literature is that international exchanges between scientists, bypassing governments and their diplomats, would be more effective in dealing with many of the global challenges that we face. There is no doubt of the global challenges we face, captured in the New International Security Agenda. There is no doubt of the need to extend debate and collaboration beyond governments and foreign ministries if we want to tackle these issues. There is no doubt either of the scientific content of many, if not all, of the issues in the New

International Security Agenda. This chapter has not questioned any of this. It has, however, sought to distinguish between the activities of science, policy advocacy and policymaking. Decisions relating to the New International Security Agenda must be informed by the latest scientific advances. Scientists must clearly be involved in ensuring that this is so. But these are not scientific decisions but policy decisions. They necessarily must involve a range of factors such as economics, geopolitics and social impact which extend beyond the science. Indeed, in some cases different areas of science may come into conflict in taking policy decisions. Scientists will take different views about the policy implications of the science (as we saw frequently during the COVID-19 pandemic). Policy decisions are frequently surrounded by considerable uncertainty and must be taken with imperfect knowledge. Historians will debate whether policymakers got it right and, even with the benefit of hindsight and the twenty–twenty vision that affords, will disagree.

The need for these decisions to be taken at an international level adds a further layer of complication and uncertainty. Once a government has decided its position it must persuade other governments, and non-state actors, that it has got it right. Here enter the science diplomats. They must not only understand the decisions their governments have taken, and the science supporting those decisions. But they must also understand the positions adopted by other governments and non-state actors, why those positions have been adopted and how those positions relate to the policies adopted by their own government. This entails the key diplomatic skill of empathy, being able to see the world and the actions of your own government through the eyes of the other. Chess players seldom flourish if they consider only their own strategies and attempt to impose them on the board without regard for what their opponent is doing. The strong chess player seeks not only to understand what his/her opponent is trying to do but also how his/her opponent interprets his/her own moves and what that means for his/her own strategy. As the Prussian Field Marshall Moltke once said, strategy is a series of expedients. The diplomat dealing with the issues of the New International Security Agenda is in a similar position. She cannot simply assume that her government has got it right, and that all other state and non-state actors are wrong, even ill-intentioned. The Rashomon effect derives from the Kurosawa film "Rashomon": three witnesses give three different but sincerely believed accounts of the murder of a samurai. In the context of international relations, it concedes that different actors can hold radically different views or interpretations of an event or a situation, all equally sincerely held. The role of the diplomat is to understand this, identify the positions and interpretations of the other actors and then identify points of synergy or shared interests that can serve as the building blocks for agreement or coalitions for convincing others. Often this implies messy compromises. But in diplomacy, often any agreement is better than none at all (which is not to say that this always is true—sometimes diplomacy is not the path to follow).

If we want to put more emphasis on diplomacy in science diplomacy, and in particular on the role of diplomats and diplomacy in tackling the policy implications of science or in creating spaces in which scientists are able to meet and collaborate uncontaminated by geopolitics, it may be worth exploring what a diplomatic approach to these issues might look like. This can also inform our discussion of cyberdiplomacy and techplomacy in future chapters. Drawing on the literature (Melissen, 1999), several elements of a diplomatic approach to international relations can be identified:

- A willingness to accept "good-enough" outcomes rather than insist on optimal solutions. Diplomats are conscious that the international problems that confront them do not have simple solutions. That is why many international issues persist through many generations of diplomats (the Balkans, for example, have been an international crisis point since the 1870s and engaged the attention of the author when he worked in the Foreign Office in the 1990s. Tensions and the risk of renewed conflict remain in the Balkans today). Diplomats are also conscious that often they are confronted by a series of bad outcomes, their ambition being limited to find and secure the least bad of them. This means that diplomats often focus on managing international problems rather than solving them. This is not from bureaucratic lethargy but from the recognition that international problems do not have solutions. As noted above, this may not be the ideal approach, especially where scientific evidence suggests that disaster looms. It may be dissatisfying for scientists and activities. But it recognizes the reality that multiple factors contribute to policy decisions and that, at an international level, other countries must be brought along by persuasion rather than force.
- As noted above, the need to bring other actors, state and non-state, for the policy decisions governments take requires an analytical approach built around identifying the intentions of "the other". This diplomatic empathy requires not only the ability to identify how other actors see the world and their intentions in that world but also how they interpret our own decisions and statements. Frequently this makes diplomats unpopular with politicians. Diplomats explaining to their political masters how other countries see the world is frequently interpreted as the diplomats supporting the views of the other countries: diplomatic sympathy rather than diplomatic empathy. Diplomats serving abroad in particular are accused of "going native". But without this ability to see the world through the eyes of the other, effective foreign policy and diplomatic strategies to implement it become impossible.
- The insights into the intentions and perceptions of other actors depend on the development of global networks of information and influence among both state and non-state actors. Networks are often posited as consequences of digital technologies, to the amusement of diplomats for whom networking has always been a key technique. Any diplomat is only as good as her contact network. But these networks are put to work not only to gather information but also to influence other actors. Through their analysis of the intentions of other actors, and how they understand their own interests, diplomats are able to identify synergies of shared interests. They can use these shared interests, or shared perceptions of interests, to construct "coalitions of the willing" in pursuit of given policy options. Traditionally, the diplomatic approach has preferred to build coalitions and alliances around shared interests rather than shared ideologies or values. The

latter excludes major countries that may not share your values and yet may help you secure a specific policy option. Dealing with the global issues of the New International Security Agenda increases the need for approaches based on interests. The Paris Accords on climate change were delivered by an agreement between Presidents Obama and Xi Jinping despite their ideological differences.
- The socialisation of state and non-state actors into an international community of actors respecting certain rules of behaviour. Historically diplomats have sought to socialize the representatives of "revolutionary" or "rogue" states thus normalizing the behaviour of these states in international relations. Thus, Soviet diplomats following the 1917 revolution sought the overthrow of the governments to which they were accredited. By the 1930s, they were part of the diplomatic community hosting and participating in cocktail parties. The Soviet Ambassador in the UK at this time, Ivan Maisky, was probably the best networked of all diplomats in London. Similar stories can be told of Chinese and Iranian diplomats following their respective revolutions. In the context of science diplomacy, diplomats seek to socialize a broad range of state and non-state actors, including scientists, into a community with agreed, albeit unwritten rules, which allows debate and coalition building around the issues of the New International Security Agenda.
- Most diplomats, especially from Anglo-Saxon countries, tend to adopt a constructivist approach to international law. International law is not obeyed because it exists or because of fear of punishment. Punishment for breaching international law generally applies only to smaller countries and their leaders (as Thucydides wrote, in the mouths of the Athenian envoys, 2400 years ago: "justice exists only between equals. Otherwise, the strong do what they will and the weak do what they can"). Rather the reasons that states and their leaders choose to obey international law result from a combination of self-interest, self-perception (and how they want to be seen by others) and a desire to remain a part of the international community.

As noted above, so far diplomats have tended to arrive late at the implications of scientific developments and new technologies for international relations, whether nuclear weapons, digital technologies and the Internet or artificial intelligence. They must therefore struggle with scientific or technological facts accomplish, where new threats to international security and stability are already well developed. This amounts often to the diplomatic equivalent of getting the toothpaste back into the foreign policy tube. As scientific and technological developments accelerate in both speed and complexity this approach will no longer be adequate. Diplomats risk getting so far behind the curve that international anarchy develops in key technological domains (as some believe has already happened in cyberspace (Riordan, 2018). This does not mean that scientists should take on more of the diplomatic role, at least not in as much as they are scientists. As we have argued, the key problems are not scientific. Rather it suggests that Science Diplomacy should also include an intensified conversation between diplomats (and other foreign policymakers) and scientists and technologists to identify the trends in scientific and technological development and what the geopolitical implications might be (this of course applies to the implications for domestic politics as well, although that is not the domain of the diplomat).

Einstein once wrote that "God doesn't play dice with the world". Kissinger wrote, slightly more long-windedly, that "The superpowers often behave like two heavily armed blind men feeling their way around a room, each believing himself in mortal peril from the other, whom he assumes to have perfect vision. Each side should know that frequently uncertainty, compromise, and incoherence are the essence of policymaking. Yet each tends to ascribe to the other a consistency, foresight, and coherence that its own experience belies". The two quotations bring out the fundamental differences between the worlds in which diplomats and scientists operate. The essence of science diplomacy is not that scientists should act as diplomats or replace diplomats. Rather it is that science is becoming increasingly central to international relations and foreign policy, either through revealing international threats requiring global responses, or through technological applications which pose challenges to international stability and security. Though driven by science, these problems in themselves are geopolitical and require diplomatic responses. Thus, in its fundament we can see science diplomacy operating in two related ways. Scientists provide the expert information that diplomats need to enable them to engage effectively with scientific agendas at an international level, and diplomats apply the skills and mindsets of diplomacy to the international problems either revealed or generated by science.

## References


Cooper, A., et al. (Eds.). (2008). *Global governance and diplomacy: Worlds apart?* Palgrave Macmillan.

Flink, T. (2021). Taking the pulse of science diplomacy and developing practices of valuation. *Science and Public Policy, 74*, 191.

Flink, T., & Schreiterer, U. (2010). Science diplomacy at the intersection of S&T policies and foreign affairs. *Science and Public Affairs, 37*, 665.

French, D. (2022). *Deterrence, coercion and appeasement: British grand strategy, 1919–1940.*

Gewen, B. (2020). *The inevitability of tragedy: Henry Kissinger and his World.* Norton 2020.

Gluckman, P. D., Turekian, V. C., Grimes, R. W., et al. (2017). Science diplomacy: A pragmatic perspective from the inside. *Science and Diplomacy, 6*, 1–13.

Gual Soler, M. (2021). Science diplomacy in Latin America and the Caribbean: Current landscape, challenges, and future perspectives. *Frontiers in Research Metrics and Analytics*. Retrieved December 12, 2021, from https://doi.org/10.3389/frma.2021.670001

Gual Soler, M. (2020). *The future of science diplomacy. Geneva: Geneva science and diplomacy anticipator (GESDA)*. Retrieved December 12, 2021, from https://gesda.global/wp-content/uploads/2020/11/GESDA-SAB-9_Future-of-Science-Diplomacy.pdf

Melissen, J. (Ed.). (1999). *Innovation in diplomatic practice*. Palgrave.

Riordan, S. (2003). *The new diplomacy, polity*.

Riordan, S. (2018). The geopolitics of cyberspace: A diplomatic perspective. *Brill Research Perspectives in Diplomacy and Foreign Policy, 3*(3), 1–84.

Riordan, S., & Torres Jarrín, M. (2020). *Science Diplomacy, Global Policy Perspective Report*. European Institute of International Studies, Salamanca-Stockholm. Retrieved December 12, 2021, from https://www.ieeiweb.eu/wp-content/uploads/2020/01/Science-Diplomacy.pdf
S4D4C Project. (2019) *The Madrid Declaration on Science Diplomacy*. Madrid. Retrieved December 12, 2021, from https://www.s4d4c.eu/s4d4c-1st-global-meeting/the-madrid-declaration-on-science-diplomacy/
Sharp, P. (2009). *Diplomatic theory of international relations*. Cambridge University Press.

# Chapter 4
# Cyberdiplomacy

**Abstract** The chapter will extend this discussion into Cyberdiplomacy. It will build on the theoretical framework built in the previous chapter to situate it in relation to science diplomacy and the broader international and geopolitical environment. It will explore the conflict of cyber agendas and the challenges of building norms of behaviour in cyberspace.

**Keywords** Cyber · Diplomacy · Cyberdiplomacy · Geopolitics and cyber regulations

## 4.1 Relations Between Science Diplomacy and Cyberdiplomacy

In the last chapter, we introduced the three axes along which diplomacy engages with other activities: agency, process and domain or subject matter. In the case of science, we were able to explore diplomatic engagement along all three axes. While we briefly discussed technology as a product of scientific activity in the context of science diplomacy as domain or subject matter, a separate area of diplomatic engagement is emerging in relation to technology and especially the digital technologies which underly the internet and cyberspace (Riordan & Torres Jarrín, 2020). Unlike Science Diplomacy, the engagement along the three axes have acquired different names. Thus, the engagement of diplomacy with digital technologies as process is normally referred to as Digital Diplomacy. In other words, Digital Diplomacy explores the ways in which diplomats can use digital technologies to promote broader foreign policy objectives or to operate more efficiently. While this is most commonly associated with public diplomacy and its use of webpages, social media platforms and search engines to influence foreign publics there are in fact a broad range of activities that digital technologies can support. Political reporting can use online media, search engines and digital networking. An innovative official at his desk in the foreign ministry can have access to sources of information at least as



M. Torres Jarrín, S. Riordan, *Science Diplomacy, Cyberdiplomacy and Techplomacy in EU-LAC Relations*, United Nations University Series on Regionalism 25, https://doi.org/10.1007/978-3-031-36868-4_4

extensive as formerly available to a diplomat posted in a capital abroad. Big data and other digital modelling techniques can support policy analysis while online simulations and gamification offer new approaches to policy making. Possibly the area of diplomatic activity which has shown the greatest innovation in the use of digital tools has been the oft forgotten consular work. Not only has the issue of visas, passports and documents been put online but the interactive use of social media platforms (e.g. during the 2015 earthquake in Nepal) has facilitated far more effective consular crisis management. There is a wide existing literature on digital diplomacy and so it will not be examined in detail in this book (Bjola and Holmes, 2015). However, it does raise the issue of what diplomats posted abroad can offer to justify the expense of their overseas existence compared to the official with a laptop in the foreign ministry. A partial answer to this question will be offered later in this chapter.

Agency explores the impact of digital technologies on the actors in international relations and diplomacy. As we have seen in Chap. 3, digital technologies have facilitated the participation in international relations of a broad range of state and non-state actors ranging from sub-national governments to corporations, NGOs and less formal civil society activists. The question arises as to which of these can be considered as diplomats or diplomatic actors. The question is not trivial. One response is to define diplomats narrowly according to the provisions of the 1961 Vienna Convention, which limits the definition to someone accredited as such to a foreign government or international organisation. This narrow definition has slightly odd implications. Diplomats working in foreign ministries, and so not accredited anywhere, are not actually diplomats until posted abroad again. Moreover, the definition of who is or is not a diplomat depends not on the government of the diplomat but rather that to which he is posted. At the other extreme of the debate, it is argued that anybody engaged in international debate is a diplomat. The trouble with this approach is that if everyone is a diplomat then no-one is a diplomat and diplomacy as a concept is hollowed out. In the last chapter, we tried to outline a "diplomatic" approach to the New International Security Agenda. If that approach has value in handling international issues, then it must be preserved which means limiting the definition of diplomat to those who share that approach. In this chapter, we will explore the value of that approach in relation to digital and other technologies (Sharp, 2009).

Another way of thinking about agency and digital technologies would be to think about the ways in which these technologies could replace diplomats. Some functions of diplomats have already been automated, as noted above. But automating certain diplomatic functions, e.g. the issue of documents, does not mean that the digital technologies used have agency as diplomats. As automated rather than autonomous systems these technologies are tightly programmed and limited in what they can do. They do not have agency as such. The challenge is to envisage technological systems acting as autonomous agents analogous to the lethal autonomous weapons systems (LAWS) being developed by various militaries: diplobots to balance the killer bots. The development of diplobots looks far more remote than that of killer bots. It is conceivable that deep learning programmes could be used in the future in complex negotiations. Each participant in the negotiations would programme its objectives and red lines into its deep learning programme. The deep learning programmes would then "compete" to produce a range of optimal outcomes. Politicians would accept or reject the outcomes (as they do when diplomats negotiate). This remains far in the future. A more immediate challenge is how to deal with the Internet and technology companies that are developing and commercialising digital and other technologies. Increasingly they seem to be shaping the international environment in ways that may qualify them as geopolitical, if not diplomatic, actors in their own right. The next chapter will explore whether governments need to develop a "techplomacy" for engaging with these companies, and what that might mean for EU-CELAC relations.

The final axis along which diplomacy engages with digital technologies is domain or subject matter (Riordan, 2019), the way in which diplomacy can be applied to cyberspace itself, or more precisely to the political and geopolitical problems that cyberspace and the Internet are generating. This is referred to as Cyberdiplomacy and is the subject of this chapter. Cyberdiplomacy is the application of diplomacy to the problems generated in cyberspace. Until recently there has been the perception that because the Internet was created by technicians, its problems are essentially technical and have technical solutions. This is reflected in the Internet Corporation for Assigned Names and Numbers (ICANN), the not-for-profit private sector corporation registered under Californian state law responsible for much internet governance. ICANN's board of directors consists of technicians, specialist lawyers and industry representatives. Government representatives, again mostly technicians, are consigned to an advisory committee. And yet the key problems of internet governance (including ICANN's status) and cybersecurity are political and geopolitical, touching on issues of sovereignty, social and foreign policy, international law and balance of power. An example of the geopolitically controversial issues with which ICANN must deal is the allocation of domains. There is debate about the allocation of the domain ".amazon". Latin American countries insist that this domain should be allocated to the peoples of the Amazon valley. A certain e-commerce company would also like to claim the domain. The allocation decision is not purely technical but deeply geopolitical. Latin American governments understandably fear that ICANN's lack of governmental representation in its executive will skew the decision towards the e-commerce company. This in turn increases the controversy about ICANN's status.

Cyberspace is developing its own geopolitics, which shape the behaviour of both state and non-state actors (Riordan, 2018). The political and geopolitical issues in cyberspace are inter-connected, and meld with parallel problems in physical space. Concerns about the global issues of the New International Security Agenda align with the issues and challenges of Internet Governance. The cybersecurity agenda, like more traditional geopolitical agendas in physical space, influences and undermines efforts to deal with the Internet Governance Agenda. While much of the debate about cybersecurity focuses on technical issues, it is driven by the same geopolitical ambitions and tensions as the wider security agenda in physical space. At the end of WWI, the French Prime Minister Georges Clemenceau is reported to have commented that "war is too important to be left to soldiers". By the same token, cyberspace is too important to be left to technicians (and intelligence services). Just as diplomacy is central to regulating and moderating the international behaviour of states in physical space, leaving cyberspace to technicians and intelligence services risks converting it into a Hobbesian space of all at war with all.

If cyberdiplomacy is the application of diplomacy to the problems arising or being generated in cyberspace it is necessary to define what we mean by Cyberspace. There are no formally agreed definitions of Cyberspace. Often it is seen as coterminous with the Internet. However, according to the 2009 British Government definition: "Cyberspace includes all forms of networked, digital activities; this includes the content of, and actions conducted through digital networks" (The UK Cybersecurity Strategy). In other words, cyberspace encompasses not only the Internet and its physical and virtual structures but also the human interactions via the Internet and other digital networks. Human interactions via the Internet are intentional and therefore the motivations for these interactions also form part of cyberspace. In other words, cyberspace extends beyond the Internet to encompass the state and non-state actors that use it, while the Internet provides the "geographical" structure underlying it.

An alternative approach is to think of Cyberspace consisting of four layers: physical, logic, data and social (Klimburg, 2017). The physical layer consists in the physical cables, transoceanic and continental, the data storage facilities of the Internet Service Providers and the Internet Exchange Points (IXPs). This physical structure is largely constructed by governments and national telecommunications companies regulated by national law, although global Internet companies are increasingly building their own infrastructure. The bandwidth and latency of the internet is dependent on the quality and structure of this physical level. We tend to think of the Internet, and cyberspace, as somehow existing in the cloud, but it is dependent on a physical structure embedded in real geography and vulnerable in the ways that all physical infrastructure is. The logic layer consists of the rules and protocols that ensure that the Internet functions and that the data arrives at its destination. It is worth noting that when the internet was being developed priority was given to ensuring the data arrives at its destination rather than to authenticating its origin. Given the context in which those communicating via the early internet were all scientists or technicians who knew each other, this is understandable. However, this means that the Internet has a built-in advantage for cyber offense over cyber defence. Concerns about ICANN's governance role, as we will see later, focus primarily on the logic layer. The data level concerns the data and information being communicated through the internet, including the content of emails and webpages. Concerns at this level focus on data protection, privacy and definitions of unacceptable content. Finally, the social level refers to the interactions between the agents using the Internet. Until recently, these agents were primarily human actors communicating with each other through the Internet. But increasingly, the social layer consists in the connection of devices through the Internet of Things.

The first three layers constitute what we call the Internet and condition the ways in which different state and non-state actors experience and understand Cyberspace, shaping geopolitical cultures and the ways these actors behave in geopolitical fields. The fourth social level is what makes the internet cyberspace. Kello (2017) has argued that we should distinguish between Cyberspace (consisting of the Internet including the World Wide Web and the cyber archipelagos) and cyberdomain, which covers human interactions in Cyberspace. But this seems an unnecessary distinction. Moreover, by distinguishing between technical and human domains, it can imply that the lower levels of the former are not political. Indeed, some commentators do distinguish between political and technical problems in Cyberspace. But, as is becoming clear, all levels in Cyberspace are highly political, and geopolitical, and politicised. These problems do not have technical solutions, as Facebook and other social media platforms have discovered when trying to tackle the problem of unacceptable content on their platforms.

Some commentators have argued that Cyberspace is flat and without structures. As a virtual world it is without borders and is able to transgress the national borders in the physical world. Khanna (2017) argues that this cross-border interconnectivity will undermine the traditional nation state, replacing it with regionally integrated coastal megacities. Others argue that direct contact between civil societies will bypass Governments, rendering them irrelevant. We have already discussed how digital technologies have facilitated the participation of a broad range of non-state actors in international debates and how the New International Security Agenda makes their participation ever more relevant and important. We have left open whether this makes them diplomatic as well as international actors. Such non-state actors include not only the familiar NGOs like Greenpeace and Amnesty International, but also companies, city and regional Governments (as Khanna identifies) and a whole host of smaller and less formal groups who are able to gather information in ways previously reserved for Governments and network at an international level. An interesting example that relates to the impact of digital technologies on governments and diplomacy is Bellingcat (Higgins, 2021). Bellingcat is a website founded by British citizen journalists which scours publicly available images and videos to uncover misbehaviour by state actors. Its successes include identifying the use of barrel bombs and cluster munitions in Syria, the Russian missile launcher which brought down MH17 over Ukraine and the identities of those suspected of poisoning the Skripals in the UK. They are part of a broader development of Open-Source Intelligence (OSINT) where investigators take advantage of the enormous quantities of data available on the Internet and social media platforms to investigate the behaviour of governments and their agents. Such techniques are reinforced by technologies such as drones and the commercialisation of satellites. Drones can access space otherwise inaccessible for security or other reasons feeding back real time video. Their low cost makes them almost disposable. Drone footage has already been used to support allegations of human rights abuses against governments. This would have been inconceivable only a few years ago.

While the Internet has facilitated the participation in international relations of this broad range of non-state actors, Khanna's declaration of the death of the nation state may be premature. Talk of the demise of the nation state was prevalent in the 1990s. Yet with the Al Qaeda attacks on New York and Washington in 2001 the nation state re-emerged as the principal defence of the physical security of its citizens. As we shall see in Chap. 6, confronted by the COVID-19 pandemic, most people looked to their national governments for salvation, even within the EU when, in principle, they should have been looking to the multinational body. In recent years, the importance of the nation state in the physical world has been strengthened by the re-appearance of more traditional international agendas centred on competition for territory, influence, and resources. Thus, Russia has effectively seized territory in Georgia as well as Ukraine, while China projects sovereignty claims into the South China Sea. A similar logic appears to function in Cyberspace. As concerns grow about the different aspects of cybersecurity, the role of Governments increases. In part this is because, as in physical space, only Governments have both the capacity and the legitimacy to undertake the full range of security measures required, including offensive operations. It is also because geography, including national borders still matter, even in cyberspace. Like physical space, cyberspace has its own geopolitics.

It is worth a quick word on what we mean by geopolitics before exploring how it can be applied to cyberspace and its implications for cyberdiplomacy. The terms "geopolitics" and "geopolitical" are now almost ubiquitous in debates about international relations. This was not true 20 years ago, when the concept of geopolitics was seen as intellectually suspect and, for some, unacceptably linked to the ideology of the Third Reich through such practitioners as Karl Haushofer (Murphy, 1997). At the same time, the high point of globalisation and America's unipolar moment (Brands, 2016) suggested that the kind of "geopolitical" agendas associated with inter-state conflict were something of the past. Thus, the return of the term "geopolitics" can to some extent be seen as a mark of disillusion with this millenarian optimism. As Kagan (2009) put it, "The world has become normal again". Geopolitics and geopolitical have become code for a realist view of the world that eschews both the liberalism and neo-conservatism that held sway either side of the millennium. Geopolitical risk has entered the everyday discourse of major corporations as well as Governments. Geopolitical risk consultancy companies have multiplied as "geopolitical uncertainty" increases, offering to anticipate international events for their clients, and protect them from the worst of their impact.

There are two problems with this sudden renewed enthusiasm for geopolitics. Firstly, it risks concealing the return of a crude form of geopolitical determinism. There is a tendency to argue that certain features of physical and political geography reflect eternal verities about the human situation, and that these predict where conflict will inevitably break out in the future, as it has in the past. The second problem is that the overextended and indiscriminate use of the terms geopolitics and geopolitical risk emptying them of meaning. As we have seen, the term "geopolitical risk" is bandied around board rooms, and consulting companies make considerable sums claiming to be able to analyse, anticipate and mitigate it. But most of the risks these consultancies analyse are not geopolitical in the strict sense of the word. Rather they are political, social, legislative or technological risks. Even risks that might reasonably be termed geopolitical—for example inter-state conflict, the imposition of sanctions regimes, trade wars—might more accurately be described as international risk. "Geopolitical risk" has, for both the consulting companies and their clients, become a convenient catch-all for all the non-commercial risks a company may confront in its international operations.

Both possibilities pose problems for applying geopolitics to cyberspace. If geopolitics is a crude geographical determinism, reflecting eternal verities about where human conflicts arise, it is difficult to see how it could be applied to Cyberspace. At most, Cyberspace could become a domain in which the inevitable conflicts arising from geopolitical verities would be fought out. But that would not amount to a geopolitics of Cyberspace. Equally, if geopolitics has been hollowed out as a concept to the extent that it is little more than a catch-all for risk analysts or a certain way of talking about international affairs, it is difficult to talk about a geopolitics of Cyberspace, other than in an equally hollowed out way. In an attempt to recover geopolitics as analytical tool that can be applied to cyberspace, this chapter will define geopolitics as the political interpretation, or the political meaning, of geography. Geography does influence the behaviour of international actors, increasing the risks of regional tensions or limiting a state's options. The point can be illustrated by the contrast between the Fulda and Suwalki Gaps. During the Cold War, the Fulda Gap was an area of low-lying land on the border between East and West Germany. Western strategists focused on it as the gap through which the Soviet tanks would thunder if the Soviet Union ever decided to invade the West. Now Fulda is a village in the middle of Germany with no significant geopolitical significance and unlikely to gain any as long as Germany remains united. The Suwalki Gap is the shortest corridor between the Russian Baltic enclave of Kaliningrad and Belarus on the border of Poland and Lithuania. During the Cold War it had no significance at all. Poland was a member of the Warsaw Pact and Lithuania and Belarus Soviet Republics. Now western strategists are concerned about the danger of Russia driving through the Suwalki Gap to join Kaliningrad with its ally Belarus while cutting off land access to the Baltic Republics. In both cases the geography remains the same but the political meaning changes.

If we want to develop a geopolitics of cyberspace, which in turn will frame the tasks and possibilities of cyberdiplomacy, we need to establish a geography of cyberspace that can be politically interpreted. One way in which the Internet functions as "geography" can be seen by examining how each of the Internet layers of Cyberspace interact with key state actors. The lowest, physical level consists of the hardware on which the Internet depends: the fibre optic cables on which data is carried, the Internet exchange points (IXPs) and the data centres of the Internet Service Providers (ISPs). Because of the way the Internet developed originally as Arpanet, part of US military research, the overwhelming majority of hardware remains within the USA. The USA has by far the largest number of IXPs (83). Because of the network package switching which underlies the operation of the Internet, and the prevalence of ISP data centres, some 80% of Internet traffic passes through fibre optic cables in US territory.

This generates some physical vulnerabilities (Sunak, 2017). A crucial underwater fibre optic cable runs from the west coast of the UK to the USA, carrying a significant percentage of global Internet traffic. British military officials are concerned about the interest Russian submarines have been showing in this cable and have dispatched British submarines to monitor their activities. It is not clear whether the Russian submarines are preparing to place taps on or cut the cables. The latter would have a dramatic effect on the operating of the global Internet. Ironically, the British know all about the importance of cutting cables: the first hostile act the UK took in the World War I was to cut the German transatlantic telegraph cable, forcing the Germans to use British cables (and allowing the British to read German diplomatic cables). More recently one of the two Internet cables to Svalbard was cut in two places. Svalbard is an important centre for satellite control. The internet cable from Hamburg to Finland was also cut following Finland's decision to join NATO. These may be accidents or politically motivated events. But they demonstrate the vulnerability of the Internet's infrastructure.

But this prevalence of physical infrastructure in the USA, and the sheer quantity of Internet traffic which passes over US controlled fibre optic cables, also shapes the ways in which other actors experience Cyberspace. This was made clear when Edward Snowden revealed the extent to which the National Security Agency (NSA) was tapping into the Internet as part of surveillance operations (Gellman, 2021). This did not necessarily mean that the NSA was reading individual emails, although it was doing that too. Without going that far, it was able to analyse the metadata, the patterns of data flows in the Internet (a technique first developed during the Second World War). The 80% of Internet data passing through the USA gave it a singular advantage. Other actors began to experience Cyberspace as being drawn into US surveillance operations. Even as they had to continue to use the Internet, they grew concerned about the security of their data, whether of individuals or Governments. In Europe, this led to proposals of a Schengen only Internet network, where data would circulate on cables physically within the Schengen zone (the borderless security zone within the European Union). China, on the other hand, is increasingly including fibre optic cables and other Internet infrastructure projects within its Belt Road Initiative (BRI) to re-balance the perceived US domination. This is sometimes referred to as the Digital Silk Road.

Hovering between the physical and logical levels of the Internet are its network structure (and that of the World Wide Web—the global network of webpages that allow most users to access the Internet). A popular theory holds that the Internet was designed to ensure the survival of US military command and control in the event of a nuclear attack. That is only partially true. Paul Baran at the Rand Organisation was asked to explore this in the early 1960s, but his proposal of (effectively) packet switching on a distributive network (with massive inbuilt redundancy) was rejected by AT&T, who then enjoyed a monopoly over US telecommunications. The packet switching was resurrected when ARPA developed the Arpanet to, initially, link universities working on defence projects. The network itself was not designed but allowed to develop as Arpanet grew and later developed into the Internet. Studies by Barabási (2003) and others have concluded that the Internet is not a distributed network but a scale free network. This means that while most nodes have very few links to other nodes, some nodes have very large numbers of links (in other words the relationship between nodes and links follows a power law distribution). This scale free network, or power law distribution, is repeated in the world wide web, where most webpages have relatively few links, but some have millions. This means that both the Internet and world wide web are more resilient than highly centralised networks but are constructed around a relatively small number of major hubs. Taking out a significant number of these hubs would not immediately take down the Internet, but the cascading effect as data had to be re-routed away from the hubs taken out would risk overwhelming other hubs eventually paralysing the flow of data. On a smaller scale this happened with the 2007 DDoS operation against Estonia. We will discuss this case in more detail later. The relevant point here is that overwhelming key government and corporate websites in a highly networked and online country crippled the country's access to the Internet and forced the government temporarily to delink the country from the global Internet to restore services. Thus the "geography" of the Internet could leave it vulnerable to a determined and targeted attack on key hubs (Klimburg, 2017).

From a geopolitical point of view, the most interesting aspect of this scale free network of the Internet and world wide web is that these hubs are geographically situated in the physical world. Although the companies associated with these major hubs, the Googles, Amazons and Facebooks, to name but three, operate around the world, their headquarters are physically centred in the USA, and primarily Silicon Valley. They see themselves as American companies and are seen as such by their users and state and non-state actors. This again shapes the experience these actors have when engaging with Cyberspace, increasing fears among some of US hegemony. But there is another aspect that reflects developments in the physical world. The concentration of hubs is progressively moving east. This can be seen in the world wide web with the exponential growth of Chinese Internet company sites like Alibaba, WeChat or Weibo. Tencent aspires to rival Google and Meta both as an owner of Internet companies and as a developer of new technologies, especially applications of AI. As China grows as regional hegemon in the physical world, it does so also in Cyberspace. The geography of the Internet is changing.

We will look at the geopolitical implications of the concentration of the major technology and Internet companies in two countries in more detail in the next chapter. It is worth, however, reflecting on the implications of the evolution of cyberspace for the continuation of the Internet as a unitary space for communications. The Internet is a single network of networks because of its history. It was developed within the USA and was released to the rest of the world at a time of maximum US hegemony. Other countries, especially authoritarian countries, may have been suspicious of the Internet and the implications for their own political stability, but they were not in a position to oppose it completely. As the Internet has evolved, countries have increasingly attempted to put constraints on it, limiting what their citizens can access. In some cases, for example efforts in Western countries to limit access to the dark web, this is aimed at eliminating criminal behaviour. In other cases, for example China's Great Firewall, governments aim to restrict the access of their citizens for political reasons and to ensure "social stability". As will see when we look at the internet governance agenda, differences are emerging between different governments over issues like privacy and protection of data. Some analysts believe that this is leading to a progressive fragmentation of the Internet.

O'Hara and Hall (2021) argue that we are seeing the emergence of four models of the internet. They label these the Silicon Valley Open Internet, the Brussels Bourgeois Internet, the DC Commercial Internet and the Beijing Paternal Internet. They also mention a fifth model, the Moscow Spoiler Model. The Silicon Valley Open Internet tries to capture the original ideals of the Internet, free and open to all with minimal government intervention. The Brussels Bourgeois Internet reflects the efforts of the European Commission to regulate the Internet to protect its citizens from its worst excesses. An example is the protection of data, in which the General Data Protection Regulation (GDPR) seeks to limit the ways in which internet companies and social media platforms can use the data collected from their customers and users. This contrasts with the DC Commercial Internet where internet data can be treated as property and monetised by its owners through targeted marketing. In exchange for their data, users receive services free at the point of delivery and monetisation incentivizes innovation. The Beijing Paternal Internet sees the internet as both a threat to public order and an opportunity to reinforce social stability. The government will control the access to the Internet of its citizens while using the collection of individual data to incentivise "good behaviour". Finally, the Moscow spoiler model is not really a constructive model of how the Internet might function. Moscow sees the Internet primarily as an opportunity to destabilise other countries. These different models of the internet are contradictory. The Brussels Bourgeois Internet is almost designed to counter the DC Commercial Internet, while companies' obligations under GDPR contradict those under the Beijing Paternal Internet. O'Hara and Hall do not go as far as suggesting that the Internet will fragment but do suggest that the growing differences between the different models will cause tensions and inefficiencies in the way it operates. A key challenge for cyberdiplomacy, both in Europe and CELAC countries, will be to minimise the consequences, economic and social, of the growing differences between the different Internet models.

## 4.2 Cyber Agenda Conflict and the Use of Cyber as a Domain for Geopolitical Rivalry

We saw in the last chapter how the conflict between the New International Security Agenda and the more traditional Geopolitical Agenda both undermined Science Diplomacy and posed its major challenge in managing the conflict. A similar conflict of agendas exists in cyberspace between the Internet Governance Agenda and the Cybersecurity Agenda. As in physical space, the Internet Governance Agenda seeks to deal with global issues affecting the regulation and use of cyberspace while at the global level the Cybersecurity agenda refers to the use of cyberspace to secure power or influence or to affect the balance of power. At a more detailed level, the Cybersecurity Agenda focuses on operations to cause physical damage, disruption, steal information or spread disinformation through cyber operations. But these are all in support of broader foreign policy aims captured by the geopolitical concepts of securing power or influencing the balance of power. Thus, the theft of intellectual property through cyberespionage may not be an end in itself but rather the means of closing a perceived technology gap that is limiting a country's foreign or security policy options.

**Internet Governance**

The data level of the Internet is the level at which the rules and protocols are agreed and operate to ensure that the data carried on the Internet gets to its destination. These rules and protocols and the way in which they are regulated grew out of how the Internet was developed by academics, engineers and computer entrepreneurs. Kello (2017) points out that the development of the TCP/-IP protocols "prioritised the successful delivery" of packets of data "over accounting of the sender's identity and their content". At the time in which Internet protocols were being designed this made sense. But it built a significant insecurity into the heart of the Internet's structure, and a significant advantage for attack over defence in cybersecurity. It is too late to change it now. The geography of the Internet matters.

International institutions tend to reflect the geopolitical power balance of the moment of their creation and be designed for the concerns of that time. As the balance of power and concerns change, the institution ceases to be fit for purpose. Thus, the UN Security Council was designed to deal with inter-state conflict, with a permanent membership reflecting the major powers in 1945 (no-one would seriously propose the UK and France as permanent members in 2023). In many respects the Internet is an international institution, and its governance structures, as we have already seen, reflect the moment and place of its birth. It was born at the moment of US hegemony, sometimes called the unipolar moment. This ensured a single global Internet. Had the Internet emerged a decade earlier, during the Cold War, it would undoubtedly not have been universal. There would have been a Soviet Internet, a NATO Internet, a Chinese Internet and perhaps a non-aligned Internet. If the Internet had emerged now, once the unipolar moment had passed, there would probably be the USA, China, Russian and EU Internets, all mutually incompatible in some respect. A recent New York Times editorial has even suggested the Internet could split into two or three parts in 10–15 years. As we have just discussed different models of the Internet are emerging undermining its coherence, Countries may be forced to choose which model to align with, which may be decided by the innovation and cost of technology being offered by each model (as we will see in the next chapter, the development of mobile network technology is being heavily influenced by geopolitical factors).

The place of the Internet's birth also shaped its regulatory framework. Silicon Valley in the 1990s still reflected a hippy-like distrust of authority and Government, and a certain entrepreneurial independence. This combination of moment and place of creation has left the Internet with a unique governance structure, largely non-Governmental and driven by engineers and technicians. As we have briefly noted before, the allocation of domain names and numbers is governed by ICANN (the

Internet Corporation for Assigned Names and Numbers), a not-for-profit company registered in the State of California. ICANN was originally contracted by the US Commerce Department to carry out the IANA (Internet Assigned Numbers Authority function) on its behalf, but this role was transferred to ICANN in its own right. ICANN's governing body consists of nominated engineers, technicians and representatives of Internet companies. Government representatives are relegated to a government advisory body. Some Governments, led by China, argue that a governance structure that was fit for purpose for networked university computers in the 1990s is no longer fit for purpose for a global public good in the twenty-first century and want Internet governance brought within an intergovernmental international organisation, possibly the existing International Telecommunications Union. Other countries, led by the USA and the EU, want to maintain the multistakeholder nature of ICANN. They have sought to argue that ICANN's role is purely technical and so should therefore be kept away from politicised organisations. These arguments have been questioned by De Nardis and Musiani (Musiani, 2016) in the context of efforts by the victims of terrorism to get ICANN to seize domain names from countries supporting terrorist groups (DeNardis & Musiani, 2016). So far, the multistakeholder approach has been able to resist the pressures from the "multilateralists", but something akin to a new Cold War is developing over Internet governance between the Free Internet Nations, which support an Internet largely untrammelled by government interference, and the Cyber Sovereignty Advocates, which want to assert national sovereignty within cyberspace. Ultimately some compromise may have to be found, possibly bringing the IANA function within the ITU or some specially designed UN body. The ILO, with its triple representation of governments, unions and employers, may be a possible model for maintaining the multistakeholder element of Internet governance.

Splits are emerging among the Free Internet Nations, reflecting evolving geopolitical cultures, especially with the election of President Trump in the USA and the progressive withdrawal of the USA from providing global public goods as part of an international liberal order. It is not clear whether the Biden administration will reverse these trends completely. A central objective of the Free Internet Nations has been to maintain net neutrality, the principle that all data being transmitted on the Internet is treated equally. Internet service providers have increasingly complained that this allows Internet media companies like YouTube or Netflix to make unfair profits while taking up an ever-greater proportion of the bandwidth. They want to be able to treat different users in different ways in transmitting their data, possibly charging companies like YouTube or Netflix a higher commercial rate (Perez & Zoraida, 2016). While this seems reasonable, Free Internet Nations and their supporters have argued that this would open up the path to privileged Internet access for wealthier Internet users, increasing the existing digital divide, and even allow some users (e.g. dissidents) to be excluded altogether. However, in June 2018 the US Federal Communications Commission voted to repeal the net neutrality provisions adopted by the Obama administration. It is yet to be seen what the international implications of this will be. The decision is being challenged in the courts, and several states, for example California, are insisting they will maintain net neutrality (although the Trump administration in turn took them to the Supreme Court for exceeding their constitutional powers). But the idea that something as fundamental as net neutrality could be decided unilaterally by the US Government on fundamentally commercial grounds will strengthen the arguments of the Cyber Sovereignty Advocates that the Internet is too important to be left to the multistakeholder model of governance and should be brought within the purview of an international organisation.

As we have seen above, a major concern of the internet governance agenda focuses on privacy and the protection of data. This has opened up further divisions between the EU and the USA. The EU's GDPR includes provisions that the data collected by a company can only be used by that company for the express purpose it was collected without the explicit agreement of the user. If implemented strictly this would undermine the business models of Facebook and other social media platforms which depend on their ability to monetise their users' data through targeted advertising. The European Commission is also taking up the issue of privacy, and in particular the right to be forgotten, whereby users can force social media platforms and search engines to eliminate permanently and completely their posts or information about them from the Internet. A further area for tension between the EU and US companies are the former's efforts to regulate online commerce, and in particular issues related to the taxation of e-commerce. Until now, US companies have proved experts in minimising the taxes they pay in the countries where they sell their goods, despite significant profits. The European Commission is discussing with European governments various kinds of digital tax which ensure that local governments also get to benefit from e-commerce within their jurisdictions. As most of the e-commerce companies affected are American, this further increases tensions between the EU and USA (O'Hara & Hall, 2021).

Other issues on the Internet governance agenda focus on international collaboration in countering criminal activities online and how to deal with unacceptable content. Diplomats are meeting in Vienna to negotiate an international convention on cybercrime. In principle international agreements on cybercrime are less controversial than some of the other issues, although disagreements about what online activities count as criminal do emerge. But discussion of unacceptable content is much more fraught. The fact that the supposedly Free Internet Countries increasingly want to keep unacceptable content off the Internet in itself shows an interesting shift towards the positions of the Cyber Sovereignty Advocates. But the definition of unacceptable content remains the sticking point. Most countries would readily agree that child pornography is unacceptable, but hate content is much harder to define. The dangers of being drawn into ever-greater censorship of the Internet are considerable.

**Cybersecurity**

If the structures of the different levels of the Internet function as the geography of Cyberspace, generating an Internet governance agenda, then the actions that state and non-state actors take against other state and non-state actors constitute the cybersecurity agenda. It is important to stress here the word "against". State and non-state actors can take many actions in Cyberspace to influence each other that may be regarded as legitimate, for example using social media platforms as part of public diplomacy campaigns. The cybersecurity agenda deals with actions that at least one of the parties regards as illegitimate. There is indeed some ambiguity here. Legitimacy to some extent lies in the eye of the beholder, and until universally accepted definitions of unacceptable behaviour in Cyberspace have been established, is likely to remain there. Some activities clearly fall within the cybersecurity agenda, for example penetrating foreign computer systems to damage those systems, cause physical (kinetic) damage in the real world or steal intellectual property or sensitive data. But other areas are more ambiguous. The difference between acceptable public diplomacy and information warfare is particularly fraught. As we will see, what we regard as acceptable public diplomacy is seen by the Russian government as unacceptable information warfare.

The problems are compounded by the tendency of the public, and an overly excitable media, to lump together a variety of different activities under the moniker cyberattacks. This is dangerous. The use of the phrase "cyber-attack" is itself misleading. Attacks in the physical world are aggressive actions that require a response, often in kind. We would not normally refer to an espionage attack. Espionage is a more or less continuous activity in which states seek to learn sensitive information about each other. States may not like being spied on by other states but accept it on the basis that they are doing likewise. The same appears to be happening in Cyberspace. This may explain the muted response to the Snowden revelations by many non-European countries. Countries like China and Russia assume the USA is carrying out cyberespionage against them: they are doing it to the USA (European outrage arose not at the revelation of US cyber espionage operations, from which they too would often benefit, but rather than the Americans were spying on their friends as well as their rivals and enemies). Referring to cyber espionage "attacks" suggest that each one must be responded to. Given the current levels of cyberespionage activities, by all sides, this is unrealistic, and risks being escalatory. As we shall see, cyber espionage may also be a crucial part of cyber defence. We should also seek to avoid the use of the term "cyberwar" (although to some extent we seem to be stuck with it). Like cyberattack, cyberwar is used to cover a multitude of sins without clarifying exactly what is meant. If we associate warfare with physical damage to persons and infrastructure, then there may only have been one cyberwar operation. If we use it to refer more widely to attacks that seek to steal data, damage information systems or plant propaganda, then we may be forced to conclude that we are already in a war in Cyberspace, and not a cold one. This is not necessarily a conclusion we would want to reach, or one that we would reach if we were more nuanced in our use of language (Libicki, 2016).

Interestingly, the question of the definition of cyberwar may be decided by the insurance industry and in the courts. NotPetya was a ransomware cyber operation targeted primarily at Ukrainian companies and organisations. However, like most ransomware operations it was indiscriminate and affected companies in other countries as well. Like most ransomware malware, NotPetya froze the data in computers so that its owners could not access it but then offered to liberate that information if the owner paid a "ransom" in cryptocurrency. A link was provided to enable the "ransom" to be paid and the key obtained to liberate the computer. Curiously, in the case of NotPetya the link did not work. The question is whether this was incompetence of the behalf of criminal hackers or deliberate. Insurance against cyberattacks is a growing area of the insurance industry. Like most insurance policies, cyber insurance policies exclude acts of war. The companies that suffered economic damage from NotPetya are claiming compensation under their insurance policies. The insurance companies are trying to decline payment on grounds that NotPetya was not a criminal act but a political act of cyberwar and is therefore excluded from the policies. The key to the case is that not functioning link. Was this incompetent criminality or a deliberate political action amounting to cyberwar? In the end the judges will decide.

Valeriano et al. (2018) distinguish between three objectives for the penetration of the computer systems of foreign states: degeneration, disruption and espionage. Degeneration refers to causing long-term or permanent damage to information systems, possibly with impact in the physical world (kinetic damage). Disruption refers to short term disruption of computer systems, which could have Kinetic effect. For example, the 2015 ransomware "Wannacry" attack, although criminal rather than political, blocked computers at some hospitals in the UK and led to the postponement of operations and other medical treatment. Espionage looks to steal data to acquire information about intentions and capabilities, intellectual property that can be used to commercial or military advantage, or materials that can be used in subsequent information operations. Libicki offers a slightly different triology: advanced persistent attack (APT), Distributed Denial of Services Attack (DDOS) and destructive attack. The problem with this trilogy, apart from its insistence in the use of the word "attack", is that it confuses method with outcome in its typology. Thus, DDOS describes a method of disruption, whereby a computer virus is used to oblige large numbers of computers to send simultaneous requests for information to a website, overwhelming its ability to respond (in this case, it is not the target, which is penetrated, but the computers unwittingly co-opted for the attack). Likewise, APT describes a long-term penetration of foreign computer systems, whether to gather information or prepare for future degenerative action. However, "destructive" describes the outcome, however it is achieved. Nye in turn distinguishes between Computer Network Exploitation (ex-filtration of confidential information against the wishes of the owner) and Computer Network Attack (using information to disrupt and destroy). But this does not really distinguish between operations aimed at permanent (including kinetic) damage and operations aimed only at short term disruption. In terms of policy responses, this is a significant distinction.

Valeriano et al. (2018) argue that states seeking to maintain power prefer degeneration operations; states rising in power prefer espionage operations; and states falling in power prefer disruption operations. They do not mean that each different kind of state exclusively uses that kind of operation, but this does seem overly determinist, much like geographic determinism in classical geopolitics. The kind of cyberoperations states or their non-state surrogates, undertake will also depend on geopolitical cultures, the geopolitical fields in which they are operating and the effect of geopolitical conditions. For example, a state might not see itself as falling in power, even if others do, and its preference for disruption operations might arise from the way it has experienced Cyberspace (including its perception of how others might operate in Cyberspace) and the influence of pre-existing military doctrines. The cyber operational decisions of a state seeking to maintain its power may be affected by a combination of national self-image, the interaction between different Government agencies, the domestic political environment the personal relationships between key political figures and relations with other countries, including allies, and the technology available.

Valeriano et al. (2018) also, wisely (see above), recommend talking about cyber coercion rather than cyber war. From the case studies they have done, they conclude that only degeneration operations seem to have compellence effect, compelling a rival to do something he did not want to. Although the only clear cut degeneration operation so far is the so-called Stuxnet attack against the Iranian nuclear programme, and it remains uncertain to what extent this really compelled Iran to change its behaviour, or led to its participation in the 2015 Joint Comprehensive Plan of Action (JCPOA). The implication may be that the level of cyber operations that have occurred so far have had a greater psychological than physical effect, generating uncertainty rather compelling behaviour. This is not to say that degeneration attacks cannot be conceived of that could have compellence effects, for example attacks with severe Kinetic effects on critical infrastructure, but as Valeriano et al. point out, for whatever reason, there have been no such attacks so far. Indeed, the vast majority of cyber operations so far have been disruption or espionage attacks.

Valeriano et al. (2018) characterise disruption attacks as essentially the weaponisation of information. They point out that even such disruption attacks can have significant Kinetic effects. The UAE sponsored penetration of computers in the official Qatar news agency in 2017, and the posting of a fake statement by the Emir of Qatar praising Iran and Hamas, criticising Trump and talking of good relations with Israel, served as a catalyst for the blockade of Qatar by Saudi Arabia and its Gulf allies. Military conflict was avoided probably only by then US Defence Secretary Mattis warning that the USA would defend its installations in Qatar. But by their characterisation of disruption as the weaponisation of information, they overlook the use of disruption operations as low-cost and low-risk weapons of punishment. The massive DDOS operations against Estonia's computer systems in 2007, whether carried out by the Russian Government or (as Russia claims) by outraged Russian patriots, aimed to punish Estonia for removing a statue commemorating Soviet war dead in the Second World War. As mentioned above, this DDOS operation demonstrated the vulnerability of scale free networks to concentrated attacks on their most important hubs. It is arguable that apparently Russian inspired attacks on Ukraine's power grid in 2015 and 2016 were also disruption operations: their effects were short lived and caused more inconvenience than real damage (they may also have been a form of signalling, to both Kyiv and Washington). Neither of these disruption operations really amounts to the weaponisation of information. And by focusing on disruption operations, Valeriano et al. may have missed the real weaponisation of information: the use of social media and other online platforms to spread dis and misinformation aimed at destabilising foreign societies and undermining and fragmenting political and social discourses, what US diplomat George Kennan would have called "Political Warfare" and the Soviets "Active Measures".

Following the 2016 Brexit referendum in the UK and the US presidential elections of the same year there has been increased focus on the use of the Internet and in particular social media platforms to carry out (dis)information operations (whether they are information or disinformation operations often lies in the eye of the beholder—the most effective information operations contain little if any disinformation as such, relying on context and plausible exaggeration to manipulate the target). There is nothing new in such operations. Kennan himself directed "political warfare" to ensure the right won the Italian elections in 1948 and the most effective "active measure", planting the story that AIDS was a CIA plot, predates the internet (and is still believed by many today, despite its disavowal by Gorbachev) (Rid, 2020). However, the Internet has greatly expanded the scope and reach of disinformation operations. In particular, the algorithms underlying social media, designed to support targeted advertising, help ensure that disinformation and fake news gets to those predisposed to believe them. That said, it remains open whether Russian disinformation operations influenced, let alone were decisive, in either the Brexit referendum or US presidential elections. Sometimes people lose because they run bad campaigns (Singer, 2018).

Conflict, including armed conflict, in physical space is constrained by a series of rules and norms that have been codified over the years: the UN Charter, The Laws of Armed Conflict (LOAC) and international humanitarian law, as well as a series of multilateral treaties and conventions on specific aspects of conflict (e.g. Geneva Convention on prisoners of war or the ban on land mines). Such laws do not act as domestic law. Outside the UN Security Council, which seems to have returned to the blockage of the Cold War, there is little effective international sanction against states. The extent to which states, or their leaders, can be sanctioned tends to be conditioned by their size and importance in the international system. To give the full quotation from the Melian Dialogue: "Justice exists only between equals. Otherwise, the strong do what they will and the weak what they can". But in a sense, this is not how international law functions (Thucydides, 2000).

In practice states generally do not decide their foreign policy in accordance with the dictates of international law, whatever they say in public. Rather they decide their foreign policy in accordance with their perception of their interests, and then adapt it to appear as if it falls within international law. They do so partly to be seen to remain within the international community of states, and partly to fit with their self-image (albeit self-delusional) as a law-abiding and responsible state. We have already discussed this in the constructivist approach to international law by many diplomats (as opposed to international lawyers).This may appear cynical, and yet it does act as a constraint on the behaviour of even the most aggressive states, provided that they see themselves as members of the international community of states and wish to remain so. Russian behaviour in Ukraine in 2014 makes the point. Russia seized the Crimea but made sure it did so with troops not wearing insignia. Russia's official presence was deferred until it had staged a referendum on Crimea's wish to become part of Russia. Although this was transparently fraudulent in Western eyes, it allowed Putin to claim a comparison with the self-determination of Kosovo by force of Western arms (in Putin's eyes at least). Although Russia destabilised Eastern Ukraine, it did not attempt to seize Ukraine (as militarily it could have) as there was no semi-legal justification it could construct for this, and it would clearly have placed Russian outside the international community and its norms. To some extent this has been undermined by Russia's 2022 invasion of Ukraine. The dictates of international law seem to have formed no constraint in this case with the Russian government making little if any attempt to justify its actions in legal terms. This may, however, reflect a certain fragmentation of the international community. The West has sought to place Moscow beyond the pale of the international community but has received only patchy support from non-Western countries. Of 54 African Union leaders invited to a video conference with Ukrainian President Zelensky (and pushed by France and Germany) only 4 took part.

It is not clear how international law and the various rules and norms of armed contact apply to Cyberspace, if indeed they do. The UN established Group of Government Experts (GGE) appeared to agree at one point that all international law applied in Cyberspace, but both the Russians and the Chinese have subsequently rowed back from that agreement. The GGE did reach some early agreements, for example on not attacking Computer Emergency Rescue Teams (CERTs), but the process seems to have stalled and it is not clear if it will continue. The European Council's Convention on Cybercrime (the Budapest Convention) seeks to prohibit some activities in Cyberspace, including data and information theft, but China and Russia are among the major countries who have not signed it. Within NATO, the Tallinn Manual (an initiative launched after the 2007 DDOS attacks on Estonia) seeks to establish principles for dealing with conflict in Cyberspace. Among other features, it distinguishes between cyberattacks that do not amount to a use of force; cyberattacks that do amount to a use of force but not an armed attack and cyberattacks, which amount to an armed attack. But these principles apply only within NATO, and amount more to guidelines for responding to cyber operations than rules of behaviour. Its conclusion that the Stuxnet operation against Iran's nuclear processing plant at Natanz "may" have amounted to an armed attack probably reflects embarrassment at fingering its leading member. A group of cybersecurity experts meeting in Las Vegas adopted the Las Vegas rules of behaviour in Cyberspace. These include the principle that any attack in Cyberspace should also be responded to with an attack in Cyberspace, and not in physical space. But these are only suggestions and not binding. NATO's doctrine, as expounded in the Tallinn Manual, does not exclude kinetic responses to cyberattacks (Libicki, 2016).

There are features of Cyberspace that mean that principles of international law and behaviour in physical space cannot simply be transferred. This is illustrated by the problem of attribution. In inter-state conflict in physical space it is usually clear who an attacker is, i.e. which state has launched an aggressive attack against another. The USA knew who had attacked it on Pearl Harbour. Article 51 of the UN Charter which enshrines the right of self-defence against aggression assumes the straightforward identification of the aggressor. Attribution can be more complicated with attacks by non-state actors, for example terrorist attacks. However, even then terrorist attacks are usually attributed quite quickly and with considerable accuracy. Indeed, most terrorist groups are only too happy to take the credit for their attacks. The problem is more likely to be deciding between multiple claimants. Governments do sometimes get attribution of terrorist attacks wrong, for example when the Spanish Government initially attributed the Atocha train bombings in 2004 to the Basque terrorist group ETA rather than an Islamist group, but these errors are usually quickly corrected. The US authorities were swiftly and accurately able to identify Al Qaeda's responsibility for the 9/11 attacks.

Attribution in Cyberspace is much harder. Hackers, whether state or non-state, often seek to conceal their identity. They have a range of techniques to use to achieve this, ranging from virtual private networks (VPNs), operating through or from third countries or using viruses to hijack other parties' computers. Governments make seek to distance themselves by working through non-state proxies, whether "patriotic" or criminal hackers. Cybersecurity experts have a range of techniques at their disposal to attribute cyber penetrations and claim that they are improving in both speed and accuracy. This has been shown by the ability of the USA to attribute attacks not only to specific countries, but also specific intelligence officers, who have then been indicted, in both China and Russia. However, attribution remains difficult and time consuming. Initial attributions may prove inaccurate and need to be corrected later. The Ransomware attack Wannacry, although criminal rather than political, was initially attributed to Russian hackers. Only later was it concluded that North Korean hackers were the most likely culprits. In the case of a serious cyberattack with Kinetic effects a government needs the capability to respond quickly, both to reassure its citizens and to deter further attacks. Identifying the source of the attack over a 6-month period with a degree of certainty of 70% is not good enough. An additional complication is false flag operations, in which the hacker deliberately makes the hack look as if it has come from another country. This creates the danger of retaliation against a third innocent country, provoking a separate cyberconflict with that country. Technical solutions may further improve the speed and confidence of attribution, but it is going to remain a serious problem in responding to cyber operations. As with other areas off cyberspace this suggests that technical solutions alone are insufficient (although they me necessary). Forensic analysis of cyberattacks, especially if state originated or sponsored, need to be accompanied by political analysis of state intentions (Buchanan, 2020).

The cybersecurity dilemma (Buchanan, 2017) offers another example where the accurate identification of intentions is crucial to mitigating conflict in cyberspace. Buchanan has explored how the classical security dilemma functions in Cyberspace and has concluded that it is even more difficult to manage than in physical space. In the classical security dilemma, country A fears the military superiority of country B. It embarks on a programme of arms purchases and other measures to improve its security situation. Country B interprets these moves as hostile and embarks on own its programme of arms purchases and security enhancements. Country A ends up worse off than it was in the first place. The classic example is the Anglo-German arms race before World War I. In the Cyberspace version of the security dilemma, country A is concerned that country B is about to penetrate its computer systems in a degeneration attack. Country A looks to enhance its security by penetrating the computer systems of country B to better understand its intentions and capabilities. Country B interprets Country A's move as a hostile preparation for future degeneration attacks and responds by increasing its penetration of country B's computer systems. Again, Country A ends up worse off.

The problem in the classical security dilemma is that country B interprets country A's moves to improve its security as hostile moves against country B, rather than defensive measures. There is a misinterpretation of what is going on. The cybersecurity dilemma is worse, in the sense that there is no way of distinguishing between a defensive penetration to ascertain capabilities and intentions, and an offensive penetration aimed at preparing for a future degeneration operation, with or without kinetic affects. Coding is just coding until it is activated, by when it may be too late. The USA, for example is aware that Russia has penetrated the computer systems operating its civilian critical infrastructure but does not know if these are essentially surveillance or espionage operations, preparations for future destructive attacks with Kinetic effects, or signalling of what the Russians can do (referred to as pwnage in the hacking world). The nature of Cyberspace also reduces or abolishes the various options for mitigating the security dilemma in physical space. In physical space it is possible to distinguish between offensive and defensive weapons (e.g. fighter aircraft are essentially defensive whereas bomber aircraft are clearly offensive). The type of weapons that country A is acquiring can help signal to country B whether its intentions are defensive or offensive. Likewise, the geographical location of new weapons systems, close to borders or in more reserved situations can also signal offensive or defensive intentions. Neither of these mitigating factors apply in Cyberspace. Accurate identification of a country's intentions and motivations is again important.

This may offer part of the answer to a question we posed at the beginning of the chapter: what value does a diplomat posted abroad add in a world of digital technologies that will justify the expense of maintaining her overseas. Both the problems of attribution and the dangers of the cybersecurity dilemma underline the importance of identifying accurately the intentions and motivations of foreign governments. If the coding of "cyberweapons", unlike weapon systems in the physical world, offer no clue to a government's intentions we need to find other ways of ascertaining them. If the true intention of coding is known only when the coding is activated, when it may be too late to respond effectively, the temptation will be to pre-empt that activation in the cyber version of a first strike. Recent studies in social neuroscience have discovered that humans are actually very good at accurately identifying intentions provided they enjoy regular face-to-face contact (and as long as they are not narcissists, which may account for the failure of the Trump-Kim summits). In other words, accurate identification of the intentions and motivations of other states requires regular face-to-face contact with political leaders, military officers, policy advisers and opinion formers in that country. Only diplomats posted in a country have access to this regular face-to-face contact (Holmes, 2018). Ironically, the nature of cyberspace, and particular coding, which in other senses threatens the role of the diplomat posted abroad, makes her more important. The accurate identification of other states' intentions and motivations becomes a necessary corollary to technical forensics in cybersecurity (Riordan, 2019).

## 4.3 Cyberdiplomacy in the Broader International/ Geopolitical Environment

Diplomats and diplomacy need to engage more deeply in Cyberspace in two senses: negotiating norms of behaviour and rules of the game with multiple state and non-state stakeholders, both in the context of Internet governance and cybersecurity; and in identifying the intentions of state actors in support of greater stability and mitigating the risks of escalation arising from the cybersecurity dilemma and the challenges of attribution. A further area in which diplomats will play a crucial role in cyberspace, as they do in physical space, is arms control. A further consequence of the difficulty in distinguishing between different types of cyber weapon, or more precisely deciding whether coding is a weapon, or an espionage tool is the difficulty in thinking about or dealing with the concept of arms limitation in cyberspace. In physical space arms limitation or disarmament talks have focused on specific types of weapons and sought to restrict their numbers or use, or on occasion prohibit their use altogether (e.g. landmines, or chemical or biological weapons). This approach makes little sense in Cyberspace. The challenges of distinguishing between different pieces of coding make weapons as such difficult if not impossible to identify. If the purpose of coding remains obscure until it is activated, it would be difficult to ban such coding in advance. Zero-day exploits, for example which are crucial to many cyber offence operations (including the Stuxnet operation against Iran) are flaws in computer operating systems. It would be unrealistic to ban them as such. Getting governments to inform the operating system companies of any flaws they discover may be more feasible from a cyber point of view but is probably politically unrealistic and would not really amount to arms control. Moreover, many of the ways of penetrating computer systems depend on manipulating human behaviour rather than technical attack. An international agreement on banning spear phishing seems unlikely. In the case of DDOS operations, the actual "attack" is carried out by innocent third parties. The uncertainty over the nature of "cyberweapons" will make states reluctant to enter agreements about them. Even if agreement could be reached on definitions of the different kinds of weapons, the attribution issue and the large-scale presence of non-state actors would make it impossible to police. A more profitable approach may be to focus on the targets of cyberattacks, negotiating at international level certain categories of target against which attacks should not deliberately be launched. As Nye (2016/2017) puts it, the taboo may "not be against types of weapon systems, but against certain types of targets". There is already precedent in physical space with states prohibited from deliberately targeting civilians (as much honoured in the breach as in the observance). A similar approach in cyberspace might focus on excluding civilian critical infrastructure (hospitals, emergency communications systems, air traffic control systems) from deliberate cyberattack. It would not be easy. There are definitional problems and dual usage issues (e.g. communications systems and air traffic control systems used both by civilians and the military). But it would offer a space in which states could engage in exchanges about cybersecurity and cyber operations with the possibility of international agreements and conventions emerging in the future.

It is clear that cybersecurity issues should move to the centre of foreign policy, rather than being left to national defence or security. Diplomats can bring a range of techniques and ways of seeing the world that can help bring greater stability to Cyberspace. As we have suggested before, leaving cybersecurity to technicians, the military and intelligence services risks escalation in intelligence operations, especially as the latter two groups will tend to see solutions in terms of responses in Cyberspace rather than conversations in physical space. As the old saw has it, if your only tool is a hammer, everything looks like a nail. Valeriano et al. argue that there have been few degeneration attacks with Kinetic effect (one definite and one possible) and that the risks of escalation may have been exaggerated. But the use of Cyberspace for degeneration, disruption and information operations is at an early stage. It is too early to assess the risks of escalation, especially, as Buchanan argues, cyber operations pose particular problems of intention identification and attribution. Diplomats are needed to help manage this and other issues emerging in Cyberspace.

This raises the issue of what diplomacy and diplomats can bring to the party. We have already seen the importance of diplomatic empathy, the ability to see a problem through the eyes of the other, or even a rival. This kind of diplomacy used to be central to the work of political officer diplomats, as shown by Kennan's famous long telegram from the US embassy in Moscow in 1946, which helped shape the US' containment strategy during the Cold War. The author's own experiences in the British Embassy in Beijing trying to understand how the Chinese government viewed the negotiations on Hong Kong, and in particular specific British proposals offers another example. Of late this seems to have been downgraded as a diplomatic skill, possibly as western governments convinced themselves of the end of history. But diplomats need to recover this capability for engaging with Cyberspace (not to mention physical space). It will be key not only in anticipating the behaviour of major players in Cyberspace, but also in identifying areas of possible cooperation, or common areas on which future agreements can be constructed.

Sharp (2009) argues that diplomats form an international community of diplomats, with common norms, attitudes and ways of seeing the world. In part, this reflects the alienation of diplomats both from the citizens of the country where they are posted and from their own fellow nationals. In some senses diplomats have more in common with each other than with their fellow nationals. This commonality of experience facilitates the ability of diplomats to manage international affairs. They can, example, keep the conversation going with diplomats of other countries in circumstances (e.g. conflict) where politicians cannot. As we have seen, Sharp highlight the important role of socialisation into this international community of diplomats. A challenge for diplomats will be to construct an international community of diplomats in Cyberspace promoting stability and managing cybersecurity issues. It may be objected that this will not be able to include non-state actors. However, this is true also of physical space, where it is difficult to conceive of ISIS ever being socialised into the international community. Good non-state actors are, however, brought within the broader international community. But this objection may be less important than it appears. The image of the lone hacker creating severe kinetic damage by hacking into critical infrastructure may be more illusion than reality. All the cyber degeneration operations and the vast majority of cyber disruption, cyberespionage and cyber disinformation operations have been carried out by Government agencies or surrogates acting on their instigation (whether "patriotic hackers" or criminal hacking groups). Indeed, the close relations between the Russian Government and organised cyber-criminal gangs (the "siloviki-organised crime symbiosis") means that socialising Russia within an international cyber community (if possible) could bring additional benefits in fighting cybercrime.

Diplomats take a variety of approaches to international law, ranging from the highly juridical approach of some European diplomats, via the more pragmatic view of Anglo-Saxon diplomats to the rejection of existing international law as a hypocritical Western imposition by some Asian countries (and on occasion Russia). There is no clear position on whether existing international law applies to Cyberspace. The EU affirms in its Cyberstrategy that it does. But both Russia and China have rowed back from apparent commitments to the same effect given in early meetings of the UN Group of Government Experts (GGE). In fact, existing international law, designed for physical space, does not translate easily into Cyberspace. Libicki (2016) points out that the concept of neutrality is problematic in Cyberspace. Negotiations to limit weapon types or numbers make little sense in Cyberspace, whereas negotiations to limit targets, especially for kinetic attack, do (e.g. putting critical civilian infrastructure off limits). Critics may argue that establishing international norms of behaviour in Cyberspace will have little value, given the ease of concealment and the problems of attribution. But they may be missing the point. International Law in physical space is as much honoured in the breach as in the observance. The English school of international relations argues that this is less important than a country's image of itself as a member of the international community and an observer of international law (the same point has been made by Nye (2016/7) about normative taboos in the context of cyber deterrence). We have argued that in any given situation governments do not ask themselves what are their options under international law. Rather they decide on a course of action in accordance with their interests, and then ask how they can make this course of action appear to be in accordance with international law. This can be an effective constraint on international behaviour because there are some courses of action that cannot be made to appear to be in accordance with international law and are therefore ruled out for countries which want to remain members of the international community (although there is unsurprising scepticism about this following the Russian invasion of Ukraine). A similar approach to international law, or norms, may have to be adopted in Cyberspace, emphasising the importance of both the empathetic analysis of geopolitical conditions and culture, and the socialisation of key actors into an international community of cyber actors.

In negotiating agreements on cybersecurity and, even more so, engaging with the Internet governance agenda diplomats will have to engage with a broad range of non-state actors. This requires developing a greater capacity for multistakeholder diplomacy: the capacity to interact with a range of state and non-state actors, and to form heterogenous coalitions built on shared preferred outcomes rather than shared values or ideologies. This has already been true of the global issues or new international security agenda, where diplomats have to engage and negotiate with a broad range of companies, NGOs and civil society groups. The same will be true of the Internet governance agenda, where agreements may proceed through bottom-up global networks of state and non-state actors, along the lines of the Paris Accords on climate change. There are dangers here for diplomats and diplomacy. Just as diplomats may seek to socialise international actors into the international community of diplomats, so other non-state actors may socialise diplomats into their own world views, whether that of the NGO or the major corporation. The risk is that the diplomat who, through multistakeholder diplomacy, is socialised into the ethical world view of the NGO may no longer be able to play her role as a diplomat (Riordan, 2019).

The major novelty in Cyberspace is the extent to which multistakeholder diplomacy is relevant to the cybersecurity agenda. Although, for now, the key players in the international cybersecurity agenda continue to be states and their surrogates, there is a further category of non-Governmental actor with which diplomats will increasingly need to deal: the Internet companies, and especially those associated with search engines and social media platforms. These Internet companies increasingly appear to act as geopolitical actors in their own right and need to be treated as such. This will be explored in more detail in the next chapter.

A further implication for diplomacy relates to cyber disinformation and disruption operations. The algorithms underlying the social media on which public diplomacy has increasingly depended limit the effectiveness of that public diplomacy, while at the same time facilitating disinformation and disruption operations. Diplomats may need to revert to older forms of more targeted diplomacy. Social media have to some extent made diplomats lazy, creating the illusion that they allow diplomats' to reach out to the whole public by tweeting or posting on Facebook. In fact, they are likely to limit the public diplomacy to those who already agree with it. This is equally true with strategic communication, which amounts to one way messaging, in itself lacking credibility, being directed by the algorithms to those who already doubt the fake news or disinformation. An earlier, pre-social media, form of public diplomacy identified key opinion formers, whom it then cultivated and sought to influence as a way of influencing foreign public opinions indirectly. A key challenge for this kind of public diplomacy was finding ways of engaging with and influencing hostile opinion formers. Combating cyber disinformation and disruption operations, in a world of echo chambers reinforced by social media algorithms, will require diplomats, and policy makers, to revert to this earlier form of public diplomacy, using the targeting and cultivation of individual influential opinion formers to bridge across the different echo chambers.

A final point relates to the way in which digital technologies have accelerated the speed of events and, even more, the speed at which they are reported and debated in social media. Technologies in the pipeline such as AI, machine learning and quantum computing will only reinforce this trend. This will force diplomatic services to rethink the way in which they are structured, and in particular the way in which embassies abroad relate to their Foreign Ministries. Effective engagement with social media already requires Foreign Ministries to delegate authority to diplomats abroad to participate in social media debates—there is simply not time to refer back to headquarters for instructions (the fact that most Foreign Ministries remain reluctant to delegate such authority means that most digital diplomacy is, by definition, ineffective). But the acceleration of events and information flows may force Foreign Ministries into a broader delegation of authority to embassies and ambassadors. The point can be made in relation to social media. Effective public diplomacy through social media involves participating in online debates where there is value in doing so. But a diplomat serving in an embassy abroad who wants to respond to some comment in social media does not have time to seek authority from his foreign ministry before responding. The online debate will have moved on long before that authorisation is received and she risks being irrelevant, if not provocative. In the twentieth century, the German army sought to respond to the combination of rapidly changing battlefield situations and the fog of war by giving greater autonomy to junior officers and senior non-commissioned officers (Shamir, 2011). The commander would set out the mission and allow detailed decisions on how to achieve the mission to be made at lower levels responding to the conditions on the ground. This was known as Mission Command (Austragstaktik in German). In the world of ever faster digital technologies, diplomacy may need to develop its own version of mission command. Foreign Ministries may be limited to setting the objectives of the diplomatic mission, leaving it up to the diplomats on the ground how to achieve them. Traditionally ambassadors are described as extraordinary and plenipotentiary. They have not been plenipotentiary since the invention of the telegram. Ironically, the latest developments in communications technologies may render them plenipotentiary again.

## 4.4 EU and CELAC in Cyberspace

The EU's norm-based approach to Cyberspace has enjoyed some successes. The Budapest Convention on Cybercrime, often simply called the Budapest Convention, was signed in Budapest in 2001 and has now been ratified by 61 nations. The Convention was the first international treaty to address Internet crime by harmonising national laws, improving investigative techniques and increasing international cooperation. However, its effectiveness has been undermined by those countries, and major powers in Cyberspace, which have not signed it and which include Russia, China and India. The Global Alliance Against Child Sexual Abuse Online was launched by the European Commission, in coordination with the USA, in 2012.

It subsequently merged with the British initiative We Protect to create the We Protect Global Alliance to End Child Sexual Exploitation Online. The European Commission is pushing the General Data Protection Regulation (GDPR) as the gold standard for international data protection and hopes that future regulatory frameworks (for example on Artificial Intelligence or encryption). The focus on cybercrime, rather than the more military aspects of cybersecurity, reflects both the civilian nature of the European Commission (the more military aspects are covered by NATO's Tallinn Manual) and the normative, "post-modern" geopolitical culture of the EU.

However, the EU's approach to Cyberspace has also encountered significant setbacks. It has neither found a successful way of countering massive Chinese cyberespionage, especially in relation to the theft of intellectual property, nor has it been able to develop an effective response to Russian disruption and information operations. In the case of Russian disruption operations in Estonia in 2007, Tallinn sought support from NATO rather than the EU. But Russian interference in the Brexit referendum and attempts to interfere in French and German elections show that these are political as well as security issues. The EU has established the East Stratcom Task Force to combat Russian disinformation campaigns in the Eastern neighbourhood. However, it is not clear how effective it is, and it does not tackle Russian information operations within the EU itself. The Snowden revelations about US cyber surveillance operations, and in particular the surveillance of European citizens and politicians, undermined cooperation with the USA over cybersecurity. Above all the Snowden revelations undermined the EU's confidence in US stewardship of the Internet. This was not helped by the decision of the Trump administration to abandon the principle of net neutrality, without prior consultation with the EU. Although the EU and the USA jointly support the multistakeholder approach, their positions have never been fully aligned. The EU appears to be becoming increasingly frustrated with the USA's unilateral approach to Internet governance while even the Biden administration complains that the EU's regulatory approach disproportionately discriminates against US companies. Cooper admits that when post-modern states clash with modern states, the modern state is likely to come out on top. That may prove as true in cyber as in physical space.

The frustrations that the EU has encountered in trying to promote its vision of Cyberspace may be producing a more aggressive assertion of the EU's normative diplomacy. The General Data Protection Regulation (GDPR), which came into effect in May of 2018, seeks to protect the privacy of the data of EU citizens. But its provisions, and protections, apply to the data of European citizens wherever that data is stored in the world. Any company, including foreign companies, that want to operate in the EU must accord the data of EU citizens the protections set out in the GDPR wherever they store it. In other words, the EU is leveraging the attractiveness of its online market to establish the extraterritorial effect of the GDPR. In a sense it is classic EU normative diplomacy, but in a newly aggressive form. As the Irish Data Protection Commissioner, Helen Dixon, who takes the lead in ensuring the compliance with GDPR by the US Internet companies based there, has affirmed, the EU is setting, and will impose, the data protection and privacy standards for the world. It seems unlikely that the Chinese will accept this. The EU risks a competition to see whose online market is more attractive.

The reactions of LAC countries to the challenges of cyberspace have inevitably been less coherent than the EU. LAC countries enjoy no comparable regional body like the EU. They share the challenge of the EU is as far as they have no major technology or internet companies, at least at a global scale (a theme that will be explored in the next chapter). But reduced to acting at national level they cannot leverage their market attraction in the way that the EU has in implementing the GDPR or in pressuring big internet companies to abide by EU rules. Engagement by individual LAC countries in the cyberspace debates has been limited and there has been little evidence of an effective cyberdiplomacy. While the EU is keen to engage LAC countries in discussions about internet governance and cybersecurity, it is forced to do largely on a bilateral level. There would be significant advantages to both sides if the EU and LAC were able to collaborate more closely on, for example, internet governance as their combined markets would increase their leverage with the major Internet companies. It is also in the interest of the EU (and the USA) to limit the penetration of the Chinese digital silk road into LAC countries. But to do so the EU must move from cyber policies focused primarily on regulation within the Union to a more effective outward looking cyberdiplomacy strategy while at the same time helping LAC countries to build up their own cyberdiplomacy capacities.

## References

Barabasi, A.-L. (2003). *Linked: The new science of networks*. Perseus.

Bjola, C., & Holmes, M. (Eds.). (2015). *Digital diplomacy theory and practice*. Routledge.

Brands, H. (2016). *Making the unipolar moment: US foreign policy and the rise of the post cold war order*. Cornell.

Buchanan, B. (2017). *The cybersecurity dilemma: Network intrusions, trust and fear in the international system.* C. Hurst & Co.

Buchanan, B. (2020). *The hacker and the state: Cyberattacks and the new normal of geopolitics*.

DeNardis, L., & Musiani, F. (2016). Governance by infrastructure. In F. Musiani, D. L. Cogburn, L. DeNardis, & N. S. Levinson (Eds.), *The turn to infrastructure in internet governance. Information technology and global governance*. Palgrave Macmillan.

Gellman, B. (2021). *Dark Mirror: Edward Snowden and the American surveillance state*. Penguin Books.

Higgins, E. (2021). *We are Bellingcat: An intelligence agency for the people*. Bloomsbury.

Holmes, M. (2018). *Face-to-face diplomacy: Social neuroscience and international relations*.

Kello, L. (2017). *The virtual weapon and international order*. Yale.

Khanna, P. (2017). *Connectography: Mapping the global network revolution*. W&N.

Kagan, R. (2009). *The return of history and the end of dreams*. Vintage.

Klimburg, A. (2017). *The darkening web: The war for cyberspace*. Penguin.

Libicki, M. (2016). *Cyberspace in peace and war*. Naval Institute Press.

Murphy, D. (1997). *The heroic earth: Geopolitical thought in Weimar Germany 1918–1933*. Kent State University.

Musiani, F. (2016). Alternative technologies as alternative institutions: The case of the domain name system. In F. Musiani, D. L. Cogburn, L. DeNardis, & N. S. Levinson (Eds.), *The turn*

*to infrastructure in internet governance. Information technology and global governance.* Palgrave Macmillan.
Nye, J. (2016). Deterrence and dissuasion in cyberspace. *International Security*, 41(3) (Winter 2016/17).
O'Hara, & Hall. (2021). *Four Internets: Data, geopolitics and the governance of cyberspace.*
Rid, T. (2020). *Active measures: The secret history of disinformation and political warfare.* Farrar, Strauss and Giroux.
Perez, M. J., & Zoraida, F. B. (2016). *A level playing field for the digital ecosystem.* Telefónica.
Riordan, S. (2018). The geopolitics of cyberspace: A diplomatic perspective. *Brill Research Perspectives in Diplomacy and Foreign. Policy, 3*(3), 1–84.
Riordan, S. (2019). Cyberdiplomacy: Managing security and governance online. *Polity.*
Riordan, S., & Torres Jarrín, M. (2020). *Cyberdiplomacy, Global Policy Perspective Report, European Institute of International Studies.* Salamanca-Stockholm. Retrieved December 12, 2021, from https://www.ieeiweb.eu/wp-content/uploads/2020/01/Cyberdiplomacy_enero.pdf
Shamir, E. (2011). *Transforming command: The pursuit of mission command in the US.* British and Israeli Armies.
Sharp, P. (2009). *Diplomatic theory of diplomatic relations.* Cambridge University Press.
Singer, P. (2018). *Like war: The Weaponisation of social media.* Eamon Dolan/Houghton Mifflin Harcourt.
Sunak, R. (2017). *Undersea cables: Indispensable, insecure, policy exchange.*
Thucydides. (2000). *The history of the Peloponnesian war.* Penguin.
Valeriano, B., Jensen, B., & Maness, R. (2018). *Cyber strategy: The evolving character of power and coercion.* Oxford University Press.

# Chapter 5
# Techplomacy

**Abstract** The chapter will extend the discussion into Techplomacy. It will explore the increasing use of technology as a domain for geopolitical confrontation and the extent to which major technology companies are emerging as geopolitical actors in their own right. It will investigate the challenges posed for governments in engaging with the major technology companies and evaluate the decision of the Danish Government to appoint an ambassador to the technology sector. It will compare the situations of both CELAC and the EU in relation to major technology and Internet companies, noting the similarities and differences and the areas of possible collaboration.

**Keywords** Techplomacy · Techambassador · Tech embassy · Big Tech Companies · Geopolitical actors and cyber regulation

## 5.1 Techplomacy and Agency in Cyberspace

Previous chapters have explored the axes of agency, process and domain or subject matter in relation to Science Diplomacy and Cyberdiplomacy. Focusing more narrowly on cyberspace, the last chapter distinguished between Digital Diplomacy, focusing on process (how diplomats make use of digital technologies), and Cyberdiplomacy, focusing on domain or subject matter (how diplomacy can be applied to cyberspace). It considered agency both in the sense of the participation of non-state actors in debates on digital technologies and the internet governance and cybersecurity agendas and of the scope for autonomous commuter systems to act as diplomatic agents. It also flagged technology and Internet companies as a special case. Techplomacy focuses on how diplomacy engages with the technology sector and the big technology and Internet companies. In part, this reflects the extent to which technology has become a battleground for geopolitical rivalry. But it also explores the agency of these companies as far as it considers the extent to which the



M. Torres Jarrín, S. Riordan, *Science Diplomacy, Cyberdiplomacy and Techplomacy in EU-LAC Relations*, United Nations University Series on Regionalism 25, https://doi.org/10.1007/978-3-031-36868-4_5

big technology and Internet companies should be considered as geopolitical actors in their own right (Riordan & Torres Jarrín, 2020).

It is worth defining what is meant by technology and internet companies. These include:

- Companies designing and producing digital hardware like Apple and Huawei.
- Companies designing and producing digital software like Microsoft.
- Major e-commerce companies like Amazon and Alibaba.
- Search engines like Google and Baidu.
- Digital and AI research conglomerates like Alphabet and Tencent.
- Social media platforms like Meta/Facebook and Weibo.
- Cybersecurity companies like Norton, Kaspersky and 360.

Some of these companies operate in more than one category and Google is of course a subsidiary of Alphabet. But this list identifies the kind of companies we will be discussing in this chapter.

Governments locked in geopolitical rivalry are increasingly using technology (Lee & Chau, 2017), and particularly digital and related technologies like artificial intelligence (AI), as battlefields through which to assert their superiority or undermine the capacities of their rivals. One aspect of techplomacy explores how their rivalries can be managed and mitigated to avoid fragmentation of key technologies and the economic damage this would cause, especially to countries in the global south.

At the same time, given the growing dependence of modern societies on digital technologies, the major technology and Internet companies shape the geopolitical environment in the way that the major oil companies did in the past. Indeed, the power and economic muscle of the major Internet companies parallels that of an oil company like Standard Oil at the beginning of the twentieth century. The US government responded to the economic and geopolitical influence this gave Standard Oil by using anti-trust legislation to break it up. Although both the US government and the European Commission have used anti-competition legislation against companies like Microsoft and Apple they have not yet taken as drastic action against major Internet companies. Conversely, the Chinese government has forced Alibaba to split its financial services from its e-commerce business. As is clear from the list of Internet companies above, the vast majority of major Internet companies are concentrated in only two countries, China and the USA. These companies include leaders in hardware and software design, social media platforms, search engines, online retailers, mobile communications technologies and AI applications. In part this results from China's decision to limit, and in some cases block, the access of US Internet companies, and in particular US social media platforms and search engines to the Chinese market. This created the space in which domestic versions of the US companies could emerge and thrive with the encouragement of the Chinese government. The EU, reflecting its regulatory geopolitical culture, preferred to try to regulate the activities of US Internet companies while allowing them to operate. LAC countries did not even do this, keeping their markets open. The sheer size of the US Internet companies and the consequent economies of scale meant that there was little chance of domestic rivals emerging. Apart from companies like Siemens, Nokia and Ericsson, all of which are limited in scale and have trouble competing with the USA and, increasingly, Chinese rivals. Efforts by the European Commission to talent spot and fund potential European Internet giants of the future have so far proved unsuccessful. LAC countries are in an even worse position. This concentration of Internet companies in two countries does not only have economic consequences. To some extent it shapes the geopolitical environment of the twenty-first century (O'Hara & Hall, 2021).

This provides second and third themes for techplomacy: how Internet companies become geopolitical actors in their own right and what this means for diplomacy and in particular how countries can develop ways of engaging with Internet companies. In debates about Internet governance both China and the USA have an interest in protecting and promoting the interests of their companies, potentially distorting the debates. The technology and Internet companies themselves participate in these debates. They are not the only non-state actors but their size and centrality to the digital economy makes them particularly influential voices. Without significant Internet companies of their own, both the EU and LAC lose influence in these debates. Moreover, as we saw in the previous chapter, whenever the EU seeks to regulate key Internet-related sectors, for example regulating the use that social media platforms make of their users' data, they are in effect only regulating US companies (the activities of Chinese social media platforms are essentially domestic). The US government tends to see this as discrimination against US companies resulting in an increase in tensions between Washington and Brussels. Moreover, as we shall see, the Internet companies themselves, intentionally not, are increasingly behaving like geopolitical actors in their own right. Without finding ways of engaging with Internet companies directly, governments (and the European Commission) find themselves even more marginalised in the key debates surrounding new technologies (O'Hara & Hall, 2021).

## 5.2 Technology as a Battlefield for Geopolitical Conflict

Previous chapters have already referred to the return of more traditional geopolitical agendas and the revival of great power rivalry if not conflict. Despite these growing geopolitical tensions, the major powers prefer to avoid direct conflict among themselves. The Russian invasion of Ukraine may appear to give the lie to this. But prior to the invasion, Russia did not regard Ukraine as a major power and specifically referred to the invasion as a special military action rather than war. In fact, the way in which the USA and other NATO countries have determinedly sought to avoid direct conflict with Russia while providing Ukraine with military equipment and training, reinforces the point. Major powers seek to avoid direct military conflict in part because of the uncertainty involved (a hard lesson that Russia is learning in Ukraine) and in part because of the risks of escalation, including the level of using nuclear weapons (a major reason the USA and its allies seek to avoid direct conflict with Russia in Ukraine). They, therefore, look for other non-military battlefields on which to play out their rivalry with less risks of escalation or at least where the escalatory ladder is easier to control. Such alternative non-military battlefields can include trade, finance and information. Even migrants can be weaponised to impose non-military coercion on other countries. The "trade war" which President Trump launched against China is a clear example of conflict by other means. The USA, concerned both by China's economic rise and the unfair trading methods it was using against the USA, imposed a series of sanctions and tariffs on Chinese exports to the USA. China retaliated with tariffs on US exports to China. Although both sides stepped back from a full trade war, tariffs and restrictions on trade with China continue under the Biden administration. The risks of escalation in a conflict fought out over international trade are less and the escalation ladder is easier to control.

Given the importance of digital technologies and the Internet to modern society it was perhaps inevitable that geopolitical conflict would spread to the technology and Internet sectors. Major powers seek to demonstrate their superiority through their domination of digital and associated technologies. As long ago as 2017 President Putin affirmed that "whoever becomes the leader in this sphere (AI) will become the ruler of the world". Technological prowess becomes a measure of a country's geopolitical power both symbolically and to the extent that it reinforces the country's economic, military and cyber capabilities. Following the confrontation between China and the US and the realisation during the COVID-19 pandemic of the vulnerability of supply chains, an element of autarky entered the calculations of the major powers. The Biden administration has sought to encourage the domestic technology sector (for example promoting the domestic production of computer chips through the so-called "Chip Act") while banning the sale of chips and other sensitive components to China. In turn, China through its "made in China by 2025" seeks to be self-sufficient in key technologies. The previous chapter explored how states use cyberspace to carry out degradation, disinformation, espionage and disinformation operations. In those cases, cyberspace is simply the domain in which states engage in rivalry and conflict with each other. But increasingly they also use the technologies on which cyberspace and the Internet are constructed and the technology and Internet companies that develop them in pursuit of wider geopolitical objectives (Miller, 2022).

The rollout of the 5G networks for mobile telecommunications, and in particular the role of the Chinese company Huawei in that rollout, provide an example of how technology and technology companies can become involved in geopolitical rivalries and the potential dangers for the global technological environment. The media controversy has centred around security concerns, and the company's relationship with the Chinese government and security services. The US government has argued that Huawei's role in the construction of the mobile network infrastructure for 5G, and associated cloud and other associated smart technologies, will give it access to sensitive data which it will share with the Chinese government. In support of this argument, the USA points out that the founder and CEO of Huawei, Ren Zhengfei, had served in the People's Liberation Army (PLA). On this basis, the US government launched a campaign, with some success, to persuade its allies to exclude Huawei technology from their 5G mobile network infrastructure. In a possible related move, it sought the extradition of Ren's daughter, and Huawei CFO, from Canada on sanctions busting charges. Meng's subsequent detention in Canada provoked a diplomatic row between Canada and China with the retaliatory detention of Canadian citizens in China.

However, there may be a more interesting and significant long-term reason for the US objection to Huawei than security. After all, Huawei technology already exists in the mobile network technology for 3G and 4G and remains there (mobile network infrastructures are built on top of each other). This relates to international industrial standards. International industrial standards ensure the international compatibility of new technologies allowing, for example a mobile phone built and operated in Europe to function in China (and avoiding the problems of regionally incompatible DVDs and DVD readers). This is not simply a matter of convenience but essential to global economic efficiency and productivity. Traditionally international industry standards for new stages of telecommunications technologies have been set by US companies or companies based in US allies like Japan or South Korea. The meetings for setting international industrial standards are not geopolitical events. They are attended mainly by technicians and industry representatives. Government representatives tend to be from technology or industry-related ministries rather than the foreign ministry or intelligence services.

5G technologies allow much faster communication over mobile telecommunications networks (reduced latency) and much wider bandwidth. This is essential to the Internet of Things (IoT) which links not only human Internet users but also a wide range of devices, ranging from domestic appliances to cars and industrial equipment (potentially the even wider bandwidth and reduced latency of 6G will be essential to the development of the Metaverse). Huawei has provided equipment and components for previous phases of mobile telephony (3G and 4G) but always as a manufacturer of equipment and components compliant with the standards set by others. The second phase of 5G, relevant to the interconnection of devices on the Internet of Things (IoT), is the first time that significant industry standards for a new technology have been set by a company based in a US rival, namely Huawei in China. The situation is even more awkward for the USA in that there are no US companies offering a full 5G package (Cisco Systems can offer only part). The main alternative companies are Ericsson and Nokia, but these are relatively small and cannot scale up in the same way as Huawei. In any event, Ericsson and Nokia are still providing equipment using the standards and patents set by Huawei. This threatens the previous hegemony of the USA and its allies in technological standards setting. International industry standards meetings, previously rather dull events for technicians, will become new battlefields for geopolitical influence. Foreign ministry officials and intelligence officers will join the technicians. Progress is already being made in developing the next phase, 6G. Huawei holds some 40% of the patents, looking again to be a major standards setter. This will presumably again be unacceptable to the USA but this time the USA will not be caught by surprise and will try to block Huawei at the standards setting meetings (Lee & Chau, 2017).

If neither side gives way there are likely to be parallel and competing standards for 6G. The danger is that in a world already endangered by fragmenting financial and trading systems as a consequence of the war in Ukraine, mobile telecommunications networks will also fragment as incompatible standards compete with each other for the critical mass needed to dominate the market. Countries in the global south will, as always, suffer most, both through the impact of the loss of global economic efficiency and productivity and as targets for the competing telecommunications systems. Decisions over which system to adopt would have serious economic and foreign policy implications. The EU, through telecommunications companies like Ericsson and Nokia, would to some extent be protected. But LAC countries, which already have significant Chinese technological penetration, would have difficult decisions to make.

The Huawei case should also alert us to the growing importance of non-Western technology companies. Whereas the European Commission has previously focused its attention on the big US technology countries, and largely drafted its technology-related directives to regulate their activities, the new challenge will be engaging with Chinese companies. Chinese technology companies are increasingly entering EU countries. Prior to the COVID-19 pandemic, Spanish department store chain, El Corte Ingles, announced that it would accept Chinese digital payment systems like Alipay (to facilitate Chinese tourism). Unlike US technology companies, Chinese companies tend to work together, offering technology packages, for example linking smart city technologies to 5G to ultra-high voltage electricity transmission. Thus, managing the interaction with Chinese technology companies will not simply be a question of protecting data or ensuring the payment of taxes in the country where services are offered (which have been the main issues with US companies). A more strategic approach to techplomacy will be needed. New technologies, and new techplomacy challenges, will arise not just in China, but also in countries like India and in East Africa. This becomes an area in which the EU and LAC countries should be working together.

Digital currencies are another area in which geopolitical rivalries could be played out over the next decade. Digital currencies can be classified according to how they generate value:

- Cryptocurrencies: Value is created by proof of work (usually solving a cryptographic problem) and in some cases an upper limit on the number of coins that can be generated. Examples include Bitcoin and Ethereum, although Ethereum has recently switched to establishing value through proof of stake.
- Private Sector Digital Currencies: Value is created by a basket of currencies and assets. This was the original idea behind Meta's (formerly Facebook's) diem, although that has now been linked simply to the US dollar.
- Central Bank Digital Currencies (CBDCs): The digital currency is simply a digital version of the physical currency with the same value, in effect a fiat digital currency. The only two examples in active use are the Bahama's digital Sand Dollar and China's Digital Yuan.

- IMF Digital Currency: A digital version of the Special Drawing Rights (SDRs) which are used to make loans from the IMF. The value of the physical SDR depends on a weighted basket of currencies including the US dollar, the Euro, the Chinese yuan, the British pound and the Japanese Yen. Former Governor of the Bank of England Mark Carney proposed using a digital SDR to replace the US dollar as the currency of international trade and investment.

Two conflicts related to digital currencies will shape the geopolitical environment of the twenty-first century: between governments and central governments and cryptocurrencies, and between the different CBDCs. So far cryptocurrencies have been mainly a speculative venture resulting in volatile valuations in terms of physical currencies. This has limited their use as practical currencies. However, they have been used for trade, especially in terms of digital products. They have also played an increasingly important role, especially Ethereum, in the gaming world. If cryptocurrencies did develop into viable currencies for ordinary transactions, they would pose serious challenges to governments and central banks. Publicly governments focus on the risks of money laundering, tax evasion and other financial crimes. But they would also undermine the abilities of governments to impose fiscal policy and central banks to control monetary policy. Users of such cryptocurrencies would in effect escape the control and oversight of their governments, at least in financial areas, seriously undermining the ability of these governments to function. To the extent that cryptocurrencies were used for international trade they would undermine the standing of the dollar as the global reserve currency. Governments and central banks are accordingly moving against cryptocurrencies. The People's Bank of China has moved furthest, banning cryptocurrencies from a whole range of transactions and also banning the "mining" of cryptocurrencies on Chinese territory. Western central banks are taking a regulatory path. The scale of transactions of cryptocurrencies remains small and they remain vulnerable to government action as long as they have to be converted at some point into physical currencies. But they remain a significant geopolitical challenge that will require diplomacy at a global level. The main challenges for cryptocurrencies are the extent to which they have become vehicles for speculation and the difficulties they have in scaling up as major trading currencies. The main vulnerability of cryptocurrencies to governmental or central bank action is the point at which they have to be converted into non-cryptocurrencies for real-world transactions. If cryptocurrencies can develop a financial ecosystem where conversion into non-cryptocurrencies is no longer necessary, even for real-world transactions, they will significantly reduce their vulnerabilities.

So far only China and the Bahamas have created their own digital currencies (El Salvador has adopted Bitcoin as an official currency, largely to reduce the commissions on the remittances on which its economy depends). The European Central Bank (ECB), the Bank of England and the US Federal Reserve are considering developing their own, although it is likely to be at least 5 or 6 years before they are introduced. Initially, it was speculated that this might produce global cooperation between central banks. The current geopolitical climate seems more likely to generate competition between CBDCs for the share of global trade. We have already noted the fragmentation of international trade and financial systems that the Ukraine war seems to have generated as countries seek to evade Western sanctions to continue trading with Russia. To avoid secondary US sanctions, countries avoid US dollars in their trade with Russia. This does not yet affect the global reserve currency status of the dollar, for which demand remains high. But an increasing proportion of trade is being done in non-dollar currencies. As China rolls out the digital Yuan at an international level it will undoubtedly try to secure some of this trade, building on the increased speed and reduced cost of transactions a digital currency can offer, especially for trade within the Belt Road Initiative (BRI). In this context, it will be interesting to see the role of digital currencies, especially CBDCs and the extent to which China will secure a first mover's advantage by offering to trade in digital yuan some 5 or 6 years before the Americans or Europeans are in a position to do so. In all events, diplomacy will need to manage the conflicts with cryptocurrencies, and to negotiate global regulatory frameworks, and between CBDCs. Facebook/Meta's diem currency has turned out to be something of a wet squib but another technology company with a greater understanding of the geopolitical challenges and implications may try again creating another agenda item for diplomats in their engagement with technology and Internet companies.

## 5.3 Internet Companies as Geopolitical Actors

It may seem strange to talk of Internet companies as geopolitical actors. Companies like Facebook/Meta and Alphabet (parent of Google) publicly eschew any geopolitical role. But there is nothing new in the geopolitics of companies. From the seventeenth to nineteenth centuries, the British East India Company was a major geopolitical actor, possessed of its own armies and navies and administering India as a British colony. In the twentieth century the "seven sisters", the major Western oil companies, were major geopolitical actors, effectively running countries and enjoying the military and political support of their home countries. It can even be argued that major US fruit companies served as geopolitical actors in Latin American politics. They certainly helped drive US foreign policy towards several Central and Latin American countries. Given the development and importance of Cyberspace, it should not therefore be surprising that Internet companies should become the twenty-first century equivalents of the "seven sisters", even if that is not how the companies see it themselves. This section will explore two kinds of Internet companies: search engines like Google; and social media platforms like Twitter, Facebook and YouTube (Zuboff, 2019). This is not to underestimate the importance of other Internet companies like Amazon and Apple. But these two categories play a particular role in the conflict in Cyberspace.

One of the keys to the geopolitical role of search engines and social media platforms in Cyberspace is the algorithms that underlie them (O'Neil, 2016). The algorithms play different roles, but in both categories they shape the geography of

Cyberspace for certain activities, both in the sense of deciding what data arrives where and in placing limits on the reach of individual actors. In the case of search engines like Google, the algorithms decide in what order the results of an individual search are presented to the user. Given that most users do not look beyond the first page of Google results, this is important if a web content designer wants to secure a high profile for their content. By studying how Google operates, it is possible to identify key criteria which decide which webpages will get priority in a search. These are known as Search Engine Optimisation (SEO) techniques. They are invaluable for e-commerce companies, journalists or politicians who want to ensure that their web content secures their profile and makes an impact (you can of course buy advertising which ensures priority as well, but the indication that your content is an advert undermines its credibility).

But bad actors can use the same SEO techniques to ensure profile is given in searches to disinformation or provocative content. For example, in late 2016 a search on Google for "Did the holocaust happen?" generated a series of webpages by holocaust denial and neo-Nazi groups. The groups had studied Google's algorithms and manipulated them to secure priority for their holocaust denial material. It took Google some time to take down the holocaust denial sites, in part because it was reluctant to alter its search algorithms (or give away too many secrets about them). This case may have produced more embarrassment than real damage, but it demonstrates the scope for malicious actors to use search engines to give profile to disinformation in an attempt to distort public debate and destabilise democratic societies. The danger is even greater with the development of "Deepfake" techniques, in which a combination of machine learning and video techniques are used to generate videos in which people appear credibly to be saying things that they have not said (and would not say). While these techniques are still at an early stage of development, and some of the fake videos produced are relatively easy to spot, they are improving rapidly. The dangers are not hard to foresee. The UAE orchestrated hack of Qatar Government systems in 2017 resulted in the posting of inflammatory comments attributed to the Emir of Qatar. These "fake statements" were sufficient to justify a Saudi-led blockade of Qatar (and secure the support of President Trump). Had the UAE orchestrated hackers have been able also to post a "deepfake" video of the Emir apparently making the statements in person, the impact would have been even more immediate, and potentially more dangerous. The danger of deepfake videos is enhanced by video sharing platforms like YouTube and Vimeo. The speed with which these platforms spread content means that the damage can be done before the victims of the deepfake can mitigate the consequences. Imagine, for example the consequences of a deepfake video of a company's CEO apparently confessing that the company's data are unreliable. The speed of the circulation on video sharing platforms would damage the company's share value long before the company had realised what was happening and issued a corrective. The scope for financial market fraud would be considerable. Technologies are being developed to identify deepfake videos, but they require time that the victims do not have. Another feature of video sharing platforms is the underlying algorithms ensuring that users receive content they will like (we will explore these algorithms further below in the context of social media platforms). They will also ensure that deepfakes will be directed to those users predisposed to believe them, further complicating the task of those seeking to mitigate their effects (Riordan, 2018).

Algorithms function differently in social media platforms. There they aim at ensuring the user receives content that he or she is likely to enjoy, be pleased with or interested in. Initially, this begins by suggesting contacts or friends that are likely to appeal to the user, to help him or her extend their social network. It then extends to content, and most importantly, to adverts tailored to the interests or tastes of the user. This essentially is how the social media platforms are able to finance themselves. However, the problem is that the same algorithms that drive tailored marketing, also ensure that the user only receives the news and opinions that he or she is likely to like or agree with. Social media did not create the echo chambers which are undermining social and political narratives in Western societies, but they are reinforcing them. More seriously, like search engines, they can be used to spread disinformation and fake news, in efforts to influence and undermine western political debates (Rid, 2020).

In a sense, the role of social media algorithms is even more pernicious than that of search engines. The algorithms underlying social media do not only spread disinformation. They also ensure that it arrives in the echo chambers most likely to believe it, where it fits in with the pre-existing prejudices of the members of the echo chamber (Singer & Brooking, 2018).

Social media companies have proved reluctant to take these issues seriously. They have appointed hundreds of monitors to identify unacceptable content and remove it. Likewise, they use algorithms to identify "bad actors" which they can ban from their platforms. But these actions are taken very much after the event and after the damage has been done. In part, this reflects the commercial requirement to protect their proprietary algorithms which ensure their incomes. They are reluctant to share these with governments. Partly it reflects the way internet companies, especially Western Internet companies, see themselves. This in turn reflects their experience of the Internet (what critical geopolitics refers to as their "geopolitical condition"). The Internet already existed before the creation of Internet companies. They were not instrumental in creating the Internet but nor did they experience it as a foreign imposition or an existential threat. Rather it was something in which they developed, and without which they could not exist. Above all it was a commercial opportunity.

Since their invention, both search engines and social media platforms have gone through various phases. The first phase is what might be called the idealistic phase, captured by Google's motto "Don't be evil" (subsequently adapted during Google's restructuring as Alphabet to "Do the right thing"). Internet companies saw themselves as new kinds of entities pursuing moral paths, as forces for good in the world. As they became ever wealthier, the moral aspects were undermined. Google's initial refusal to collaborate with Chinese Government censorship had broken down by 2018, as it agreed to develop a special search machine for China (to the consternation of many of its own workers who appear to have forced Google to abandon the Dragonfly project at the end of 2018). Facebook found itself caught up in a series of scandals about the use it allowed to be made of its users' data. The Cambridge Analytica scandal revealed that Facebook had allowed Cambridge Analytica to use applications to gather Facebook users' data to allow more targeted campaigning in the US elections. Particularly controversial was a questionnaire that Cambridge Analytica developed which not only allowed the company access to the data of users who filled in the questionnaire but also of all their "friends" on Facebook. These problems and scandals marked a second phase in the Internet companies' experience of Cyberspace, where they moved from being forces for good to being seen by the public, and reluctantly by themselves, as mechanisms for monetising the data of their users. This provoked growing concerns among the public and Government, in particular in western Europe, about the use and protection of citizens' data. As noted previously, this culminated in the European Commission developing the GDPR to protect the data of European citizens. However, the Internet companies are still unwilling, or unable, to accept or assimilate the third phase of their experience of Cyberspace: as active players in geopolitical conflict in Cyberspace (Zuboff, 2019).

Given, as we have seen, the ability of the algorithms underlying social media to ensure that disinformation and fake news get to the echo chambers already predisposed to believe it, it is unsurprising that those using information warfare techniques to undermine the political and social stabilities of Western societies would take advantage of them. There is now considerable evidence that Russian operatives, working directly for Russian intelligence agencies or indirectly as controlled surrogates, made extensive use of Facebook, Twitter and other social media try to influence the 2016 Brexit vote and the US Presidential elections the same year, as well as elections in France and Germany. It is not simply that the social media platforms are being taken advantage of by countries launching disinformation and disruption operations. The very design of the algorithms underlying them gives an asymmetrical advantage to disinformation operations. Three key features of disinformation and disruption operations explain this: they are not aimed at convincing those that do not agree with them, but rather at undermining confidence in all narratives; they are not overt, in the sense that the Government behind them does not openly own up to the operations, often hiding its role behind surrogates; because of the previous two points, they do not have to be coherent, but can segment their messages to appeal to different echo chambers. The Western public diplomacy response is different: it does seek to convince people; it has to do so because it is overt, in the sense that the Governments behind it admit to it; and therefore, it must be coherent to be convincing. Disinformation and disruption operations can segment their messages and fake news to appeal to different (often opposed) echo chambers while relying on the social media algorithms to deliver them to the echo chambers predisposed to believe them. In one case, the same disinformation operation was operating a Facebook page for Black Lives Matter and another defending the police. One page would carry fake news about police violence while the other carried fake news about the "criminality" of ethnic minorities. Facebook's algorithms ensured that the different fake news got to those predisposed to believe it. But the same social media algorithms will ensure that coherent Western public diplomacy messages will reach only those who already agree with them. In short, social media algorithms facilitate disinformation and disruption operations but frustrate public diplomacy (Singer & Brooking, 2018).

However, the social media companies are reluctant to accept the extent of the problem, or the proactive role played by their algorithms. Facebook is a case in point. Reluctantly, under pressure from Western Governments and media, Facebook has agreed to take down pages thought to be used by Russian intelligence officers or bots. But it remains unwilling to share details of its algorithms with Western Governments. It portrays itself as a neutral platform that Russians and others are taking advantage of. But it may not understand the concept of neutrality in international law. Neutrality in physical space carries both privileges and obligations and cannot always be maintained. One obligation is that a neutral country shall not allow another country's armed forces to cross its territory to attack a third country. In 1914, Belgium's neutrality had been guaranteed by other European powers, including Prussia, in the 1830 treaty which had created the country. When the German Government requested passage across Belgian territory for its troops to attack France, the Belgian Government was faced with a difficult choice. If it acquiesced to the German request, it breached its neutrality obligations and became a de facto ally of Germany. If it refused the request and resisted the German advance, it became a de facto ally of France against Germany. It chose the latter and paid a terrible price. Ironically, it is difficult to see how this provision of neutrality could apply to countries in Cyberspace. As Libicki points out (Libicki, 2016), given the way that the Internet works by splitting up packets of information and re-assembling them at their destination, it is virtually impossible for Governments to be aware of all the information passing through the optic fibre cables passing through their country. It would therefore be most unreasonable to hold them responsible for data passing through the cables on their territory as part of a cyberattack on a third country. However, the provision can be applied to social media companies. Either they allow disinformation operations to continue on their platforms, in which case they become de facto allies of those carrying out the operations (most recently Russia). Or they collaborate fully with western Governments (e.g. by sharing details about their algorithms with western Governments), in which case they become de facto allies of the West against Russia and other disrupters. This latter, of course, runs a similar risk as that run by Belgium: by siding fully with the West they negate their usefulness for Russian disinformation operations and may themselves become targets for disruption and degeneration attacks.

The social media platforms are reluctant to see it like this, not least because they are reluctant to share details about their algorithms. These algorithms are their competitive advantage and drive their incomes from advertising. They are reluctant to share them with Governments, concerned that the details could leak. Thus, we have geopolitical actors who refuse to see themselves as such because of the implications for their business models. But just as they have become embroiled in arguments about data protection, they will inevitably get drawn further into debates about their role in disinformation campaigns. Even if search engines and social media platforms do not want to see themselves as geopolitical actors, they do play a key role in shaping the Cyberspace environment, and its geography, where some forms of cyberconflict take place.

Three concepts from critical geopolitics can help us analyse the position of Internet companies in cyberspace. Geopolitical culture describes the way a country sees itself in relation to the geopolitical environment and the way it thinks about engagement with other international actors. It depends in part on a country's political country and partly on its perception of its own history. As we have seen, the geopolitical condition describes how a country has experienced new technologies, in this case digital technologies. Finally, geopolitical fields describe the different virtual and physical spaces in which a country operates. For Internet companies, the geopolitical condition is the experience of digital technologies not only as something normal but as essential to their own existence. The geopolitical field in which they operate initially centred on information gathering or building social networks but increasingly this is fragmenting into different geopolitical fields describing their interactions with users, advertisers, governments and each other. Their objectives and behaviour in each of these fields can vary dramatically. Unlike states, they do not have a geopolitical culture that precedes, or is separate from, their engagement with Cyberspace. Their geopolitical culture derives from their engagement with Cyberspace. Thus, deriving from their initial geopolitical field, spreading information and promoting social interaction online, it is positive, seeing them as almost altruistic operators in Cyberspace. Only slowly as the geopolitical fields in which they operated shifted to commercial marketing and the monetising of their users' data did their geopolitical culture have to accept the commercial element, but still alongside their self-image as a force for good. They still have not been able to assimilate the implications of the final geopolitical field in which they operate, the battlefield for information conflict between, primarily, Russia and the West. From the point of view of geopolitical culture, they remain in denial (perhaps analogous to the kind of self-denial generated by US exceptionalism) (Riordan, 2018).

It may be worth mentioning here Zuboff's work on surveillance capitalism (Zuboff, 2019). Zuboff's thesis is that internet companies discovered that they were gathering a lot of "waste data" about their users, data that they did not need to improve the services being offered to their users. They realised that this data could be processed to construct profiles of their users which could be sold to advertisers in pursuit of the advertising grail of targeted advertising. This was supported by the algorithms we have already discussed. The key was to minimise the uncertainty in the behaviour of their users to the benefit of the advertising customers by detailed profiles of users to which advertising could be directed. This became the business model of search engines as well as social media platforms. But it is not simply a question of the data which users consciously upload onto these platforms. With the advent of smart phones enormous quantities of data can be collected and processed without the awareness of the users. In an amusing example, data collected and published by a running app identified the location of American military bases. Google will even process how users of its search engines misspell their searches in building up their profiles. This in effect undermines the effectiveness of the General Data Protection Regulation (GDPR). Zuboff argues that companies like Google and

Facebook now wish to take this a stage further by using these profiles not just to predict their users' behaviour more accurately but also to shape that behaviour by providing constant and targeted prompts to their users. At the relatively innocent level, this would amount to reminding a user about their love of strawberry donuts when passing a store selling strawberry donuts. But it is not hard to see how this use of profiles for targeted prompting of users' behaviour could have more dystopian interpretations. At the very least, it gives an indication of the power that these companies acquire through their access to vast quantities of data and the capacity to process and reinforces the suggestion that they should be treated as more than just commercial companies.

The wider activities of Facebook's parent company Meta make even clearer the extent to which it has become a geopolitical actor, whether or not it admits it to itself. We have already noted how Facebook, as it then was, planned to issue its own digital currency based on a basket of currencies and assets. The aim was to generate a currency linked to the Facebook platform which would allow anyone in the world to be able to trade internationally without the need to access US dollars or a bank account. The basket of currencies and assets would be chosen by Facebook and would make the new digital currency independent of any given physical fiat currency. Zuckerberg did not hide his ambition to replace the US dollar for many online transactions. The creation of such a currency would have been a deeply geopolitical act with implications that Facebook itself seemed to have underestimated. Apart from the dangers of money laundering and other illegal activities taking advantage of a currency unregulated by any central bank, a global digital currency of the kind to which Facebook appeared to aspire threatened to undermine the international reserve currency status of the US dollar. This reserve currency status of the dollar affords the USA significant geopolitical advantages. Firstly, it enables the USA to run fiscal deficits that would be unsustainable for other countries. But possibly more significantly, it also allows the USA to impose unilateral sanctions on countries, backed up by secondary sanctions on companies which then breach those sanctions. It is able to do so because the dollar remains the international currency of trade, and companies need to be able to trade in dollars, making them vulnerable to secondary sanctions. Zuckerberg appears not to have appreciated the geopolitical significance of launching an international digital currency or anticipate the backlash from the US Congress and the US Federal Reserve. Various sponsors, unnerved by the backlash, pulled out. The proposal has now been reduced to a digital currency based on the dollar (and therefore no threat to the global reserve status of the dollar) for use by Facebook users. Even in this modest form it has not yet emerged.

Mark Zuckerberg reinforced this sense of Facebook/Meta as a geopolitical actor when he announced Facebook's commitment to the creation of the Metaverse and in particular Facebook's investment in Augmented Reality (AR) and Virtual Reality (VR) technology and the glasses or headsets necessary to access it. This commitment to developing the Metaverse was symbolised by his decision to change the name of Facebook's holding company to Meta. Meta is not the only company gambling on the development of the Metaverse. Tim Sweeney, CEO of online gaming platform Epic Games, has "had Metaverse ambitions for a long time". Sony and

Microsoft are also committed to its development. Some commentators have argued that the Metaverse is the next logical stage in the evolution of Internet. The Internet has evolved from early modem connections through personal computers and laptops to the current mobile networking phase. Our current level of online interconnectedness through smart phones and the devices of the Internet of Things would have been inconceivable during earlier phases of the Internet, even at the time of the introduction of the first iPhone. The level of online interconnectedness of younger people, whether communicating with each other through online messaging apps or through online gaming platforms has even led some commentators to suggest that many people already live in a primitive version of the Metaverse.

Metaverse commentator Matthew Ball (Ball, 2022) defines the Metaverse as: "A massively scaled and interoperable network of real-time rendered 3D virtual worlds that can be experienced synchronously and persistently by an effective unlimited number of users with an individual sense of presence, and with continuity of data, such as identity, history, entitlements, objects, communications and payments". This is an ambitious description in which the Metaverse performs like a genuinely parallel world with the main difference being that we can choose our own avatars to express ourselves as we would like to appear rather than as we are. An idea of what this world might be like can be seen in Neal Stephenson's dystopian novel Snow Crash (Stephenson, 2011). But it requires that large numbers of people can have real-time access, that the identity of avatars is stable throughout time, that past events are remembered in the system as are entitlements (in the sense of qualities, properties or property that avatars collect), that avatars and their entitlements can move between different parts of the Metaverse (interoperability) and that 3D events can be rendered in real time. These pose significant challenges that existing technology cannot solve. Particular problems include the lack of computing power, high network latency and limited bandwidth and low-quality VR and AR equipment. Zuckerberg admits that one of the great challenges will be providing sufficient computing power in headsets that nevertheless remain wearable. Meta's proposal of including more sensors in its VR devices to ensure that avatars reflect users' real-life expressions raises significant data protection and privacy issues (Zuckerberg does not have a good record on either). Indeed, issues related to data protection, privacy and online governance promise to be even more difficult in the Metaverse than they are in cyberspace.

Significant questions remain about the Metaverse, even if the technological issues can be resolved. These include:

- Will it be constructed top-down by corporations like Meta, Microsoft or Google, or will it evolve bottom-up from the progressive combination of smaller Virtual Reality worlds and gaming platforms?
- What role will governments play in the development of the Metaverse?
- Crucially, will there be a single Metaverse, albeit a network of smaller Virtual Reality worlds, or multiple metaverses in competition or even conflict with each other?

The construction of the Metaverse implies significant challenges for diplomacy. The Metaverse will not arise as a unitary fait accompli in a world where US hegemony is sufficient to minimise state opposition like the Internet. It will be geopolitically challenged from the beginning. The European Union is already wondering how to regulate it. Authoritarian governments have a vested interest in establishing sovereignty rights and undermining a single Metaverse which might threaten regime stability. Internet and gaming corporations will need to negotiate among themselves and with governments on the terms and protocols for the creation of the Metaverse. This may cause a demand in the corporate sector for the kind of negotiating and networking skills in which diplomats specialise. It remains to be seen whether they seek to meet these needs by recruiting diplomats, and what strain this places on governments' diplomatic resources (we will see later how Microsoft was able to outbid the Danish Foreign Ministry for the services of its Techambassador). Once the Metaverse is constructed it will need rules, protocols and norms of behaviour if it is not to become a Hobbesian world of all against all. Leaving this to the corporate builders of the Metaverse risks a loss of democratic legitimacy and accountability. Many advocates of the Metaverse, like many original advocates of the internet, are sceptical of government involvement and would like to see the Metaverse free from government interference. This has not proved possible with the Internet, and it seems unlikely that some form of government involvement in the Metaverse will be avoided either. If the concept of sovereignty is not established in the Metaverse, it may make little sense having virtual embassies there other than for public diplomacy and national marketing. There is general agreement that a thriving digital economy will be essential for the sustainability of the Metaverse, probably backed by cryptocurrencies, non-fungible tokens (NFTs) and blockchain technology (Tapscott & Tapscott, 2016). But this raises the questions of who will enforce contracts and guarantee property rights (the essential roles of governments in real-world economies): governments, a new Metaverse governance body (in which case who will belong and how will they be chosen?) or decentralised blockchain technologies (as the libertarian supporters of the Metaverse advocate)?

The issue of terrorism illustrates how a real-world phenomenon could infect the Metaverse and require a diplomatic response. The Metaverse will inevitably be a tempting target for terrorists and terrorist groups (at it will for organised crime). How terrorists might take advantage of the Metaverse is suggested by terrorist activity in cyberspace. In spite of the fears about cyberterrorism, there is limited if any evidence of any groups trying to carry out direct terrorist attacks via cyberspace (although one might argue that some of the activities of state-backed non-state actors in cyberspace get close to terrorism). There may be many reasons for this. One will be the technical difficulty of carrying out a cyberattack that has the same visual impact as planting a bomb. In general, terrorist groups have preferred to use cyberspace for raising funds (whether through donations or criminal activities), communications, recruitment and propaganda.

Terrorist activity in the Metaverse may depend on whether there is a single Metaverse or competing metaverses. Cyberterrorism may be more attractive in a single Metaverse. It depends on the socio-economic importance of the Metaverse in society and the psychological dependency it engenders among its users. Terrorists or motivated groups of activists could carry out cyberattacks to delay or disrupt the construction of the Metaverse. Disinformation operations could sow dissent among the various actors in the construction. In this, they could receive support from state actors suspicious of the creation of the Metaverse. Once the Metaverse is created they could launch cyberattacks against individual avatars or groups of avatars or destroy or damage symbolic virtual institutions. This could cause economic or psychological damage with low risk for the terrorists. The ability of terrorists to raise funds through donations, selling non-fungible tokens (NFTs) or scams and other criminal activities would be facilitated by the expected reliance of the Metaverse on cryptocurrencies. Multiple metaverses in competition or conflict would create similar opportunities for terrorist groups. But it would also offer them the chance to create their own metaverses driven by their own ideologies. These could be used for communication, propaganda and financing. Even more attractive would be to attract young people to play out their fantasies in an unattributed terrorist metaverse as a first step into recruiting them for terrorist activities. Monitoring each metaverse in a world of multiple competing metaverses will be a major challenge for security forces.

The Internet took diplomats by surprise. They were slow to understand that not all problems in cyberspace were technical ones that could be left to technicians. Only relatively recently have they realised that the most pressing problems in cyberspace are political and geopolitical, and need diplomatic management. The Metaverse has been long advertised. The diplomats have no excuse not to start thinking about how to engage with it. They, and their governments, must seize the opportunity to shape it. Governments must decide what kind of Metaverse they want to see and the norms and rules underlying its construction and operation. Above all, they must seek international cooperation to limit the scope of action of terrorists whether in a single Metaverse or multiple metaverses. This will require an effective multistakeholder diplomacy in which diplomats can engage effectively with a range of non-state actors from Internet companies and gaming platforms to users and activist groups. It also requires diplomacy based on interests rather than values and ideology. Effective diplomacy to limit the activities of terrorists in the Metaverse cannot exclude major and influential countries because we disagree with their values or ideology.

The reasons why this discursion of the Metaverse is important are twofold. Firstly, it shows yet again how the major Internet companies are straying from purely commercial areas into areas more appropriate to state actors. The effort to develop a global virtual world parallel to the real world is undoubtedly a political if not geopolitical act. Its implications are highly geopolitical, as are the controversies it is likely to cause. Secondly, it again demonstrates the need for diplomats and diplomacy to engage with technology and internet and technology companies, not just as commercial operators but as fellow geopolitical actors. In the last chapter, we discussed the geopolitics of cyberspace and how this conditions the engagement of state and non-state actors in cyberspace. The geopolitics of cyberspace is still at a very early stage and still needs much work. But analysts and diplomats already need to start reflecting on the geopolitics of the Metaverse.

There is another sector of technology and Internet companies with clear geopolitical implications: private cybersecurity companies (PCCs). At the moment, PCCs focus on offering passive cyber defence for their clients and cyberattack attribution, produced by their cyber forensic capabilities (the technical analysis and tracking back of cyberattacks to their origins). Most PCCs allow their staff 10% of their time for "research", much of which is devoted to honing their skills in cyber forensics. The cyber defence and cyber forensic capabilities of these companies exceed that of most governments. Their offer of passive cyber defence services to both governments and companies is relatively uncontroversial. Their publication of cyberattack attributions is more complicated. Although some attributions are for clients, in other cases PCCs publish attributions for marketing reasons. In the case of a severe cyberattack, possibly even causing physical (Kinetic) damage or loss of life, the victim government would come under pressure to respond. It may not want to publish an early attribution, either because it is not sufficiently confident in the attribution or because it does not want to publish the attribution until it has decided how to respond. PCCs are unconcerned about such considerations. But if the PCC publishes its attribution for marketing reasons, it puts the government under enormous pressure to act. The result can be an unnecessary escalation of response or even a new cyberconflict if the attribution is wrong (the 2015 Wannacry ransomware attack was originally attributed to Russian hackers. The attribution was subsequently changed to North Korean hackers).

The position becomes more critical if PCCs develop the capacity for offensive cyber operations and offer these services to governments. There is also considerable debate about active defence measures in cybersecurity, where the hacker himself is hacked to prevent further attacks. Most governments prohibit private companies from doing this, although it is an effective form of cyber defence, because it seems too close to cyber vigilantes. Governments try to retain the monopoly on legitimate "violence in cyberspace". But with the majority of governments lacking the offensive cyber capabilities that PCCs are able to develop, it seems inevitable that at some point PCCs will offer these "cyber mercenary" services to governments (and possibly also companies). To some extent this already happens. In the US federal pay scales restrict how much government cyber agencies like the National Security Agency can pay their cyber technicians. This means they frequently lose out on recruiting the brightest technicians to the private sector. They get round this by using contractors who are not bound by federal wage scales and rules. These contractors frequently are involved in offensive operations. Edward Snowden was working for Booze Allen, not the NSA or CIA, when he leaked the documents about US cyber activities.

This use of contractors for offensive cyber operations raises similar questions to the use of Private Military Security Companies (PMSCs) like Blackwater as contractors in Iraq (Egloff, 2021). The performance of Blackwater in Iraq demonstrated problems relating to chains of command, accountability, legal responsibility and rules of procedure. Similar problems in principle could arise with PCCs operating as contractors for offensive cyber operations for governments who otherwise cannot attract the best-qualified technicians (like the USA) or who lack the offensive capabilities themselves. Russia has not yet used PCCs for offensive cyber operations, preferring to use criminal cyber gangs when they wish to use surrogates for cyberattacks. But they have used a private company for disinformation operations and spreading fake news through social media platforms, the Internet Research Agency (ironically better known by its acronym IRA) and increasingly make geopolitical use of the private military security company Wagner. Apart from these examples, the dangers of cyber mercenary PCCs offering their services to governments for offensive operations appear to lie in the future, but it may be a very near future. Any effort to regulate the performance of PCCs at a global level will require effective multilateral diplomacy, in which the interests of the EU and LAC countries should be aligned. It will also require effective engagement with the PCCs themselves.

## 5.4 The Techambassador

In 2017, Denmark appointed an ambassador to the high technology sector, commonly known as Denmark's Techambassador. France and Germany followed suit in 2018. However, the French Ambassador appears to be mainly concerned with preventing the use of social media and search machines to promote terrorism. Following a meeting with the German Envoy for Technology and Digitalisation, Denmark's first Techambassador Caspar Klynge tweeted that there was significant overlap between their roles, but it is not clear what this is. The German Envoy is not an ambassador to the technology sector as such, and his role seems more ambiguous and less innovative. Other nations have envoys responsible for the technology sector or digital technologies, but they are not configured as ambassadors as such.

The Danish Techambassador is not the first time a country has opened a diplomatic mission related to cyberspace. The Swedes famously opened an embassy in the virtual world "Second Life", although Second Life proved short lived, and the Swedish Embassy had the feel of a gimmick. Several countries, especially those with limited resources, have opened virtual embassies, which offer information and services online in foreign countries. But these are cases of diplomacy taking advantage of digital tools, either to improve reach or to save resources, (often called "Digital Diplomacy") rather than engaging with the generators of new technologies as diplomatic actors. In this sense, the Danish move is a first and innovative. There is talk of creating embassies in the Metaverse promoted by Mark Zuckerberg but it is not clear what this would mean. As suggested above, diplomacy might be more important in creating the Metaverse than participating in it.

The Office of Denmark's Techambassador (as the Tech Embassy is officially called), is not limited to one physical site (as would be the case with a bilateral embassy) (Ministry of Foreign Affairs of Denmark, 2022), but split across three sights: Silicon Valley, Copenhagen and Beijing. There is talk of opening a further office in East Africa. Denmark's first Techambassador Klynge coined the term

Techplomacy to refer to his role (and has launched a podcast "Techplomacy" to explore and promote it further). He has described the key roles of his mission as:

- Gathering information on new and future technological developments and analysing their impact on diplomacy, politics and Danish society.
- Gathering information on developments within the technological sector itself, relations between tech companies and their plans for future operations/ investments.
- Discussing ethical and regulatory issues with technological companies, with a focus on unacceptable content and data and privacy protection.
- Convincing technology companies to base operations, research facilities or European subsidiaries in Denmark.
- Promoting Denmark as a highly digitalised country at the cutting edge of new technologies and thinking (Klynge's appointment as Denmark's Techambassador itself contributes to this).

Klynge's description of his role raises several interesting issues. The different offices out of which he operates (Silicon Valley, Copenhagen and Beijing) give a fair idea of where Denmark sees as centres of technological innovation in the twenty-first century, as do the plans for a further centre in East Africa. It is interesting that they have not situated an office in the EU (other than the office in Copenhagen, which effectively gives the Techambassador a home base). But then it is also interesting that it is the Danish Foreign Ministry, rather than the European External Action Service, that has taken the initiative to set up a Mission to the Tech Sector.

Although Klynge proclaims a new kind of diplomacy, Techplomacy, for his mission, his job description looks remarkably like that of a traditional Ambassador or Embassy (Riordan & Torres Jarrín, 2020); gathering and analysing information about the country in which he is stationed and its implications for his home country; engaging the host government in discussions about regulatory issues and international law (particularly relevant in EU countries); engaging with the commercial and financial sectors to attract Foreign Direct Investment (FDI); and the promotion of his own country. The only two key activities of traditional embassies missing are engaging with the host government on geopolitical issues (but see below) and consular protection. The latter may be implicit, however, in the discussion of ethical and regulatory issues—at least in part aimed at the protection of the rights of Danish citizens in cyberspace. It is not that diplomacy is different, but that its target has changed, reflecting the development of cyberspace and the importance of technology companies. Indeed, Klynge's appointment recognises the importance of technological developments, especially in digitalisation, artificial intelligence and machine learning, to the development of international relations, and to the physical and economic security of countries like Denmark. On the Mission to the Tech Sector's webpage, Klynge explicitly recognises that "Techplomacy" should be seen as complementary to traditional diplomacy.

Klynge sought to build multistakeholder platforms with the tech sector, and other groups or individuals interested in the key issues, which can support the effort of his colleagues working with governments and international organisations. The Mission also implicitly recognised Internet and technological companies as geopolitical players in their own right, who are shaping the geopolitical environment in both physical and cyberspace, and in which states have to operate. This is not the first time that private companies have shaped the geopolitical environment—it could be argued that both telecommunication and oil companies have done so in the past. But no country in the past has felt the need to create a Mission and Ambassador specifically to these sectors. Something has changed. Much of the work Klynge appeared to be doing initially (to judge from the Mission webpage and his Twitter feed) focused on Public Diplomacy-like activities to engage with Internet and Technology companies, but also with a wider public to promote understanding of his mission. Thus, he participated in a series of conferences and workshops, including attendance at the UN General Assembly. But he launched "Ambassador for one Day", in which US university students (interestingly not Danish university students) could compete to spend one day replacing Klynge as Denmark's Techambassador. These Public Diplomacy activities no doubt helped establish the standing and credibility of Denmark's Mission to the Tech Sector, and as such contribute to the Mission objectives/job description (as set out above). However, there are other issues about how Klynge and his team related to the Tech Sector, and in particular Internet Companies (especially social media platforms and search engines).

The Danish government appointed a Techambassador to develop Denmark's relationship with the Tech sector, promote Denmark as a centre for technology innovation and ensure that Denmark remains abreast of the latest technological developments. The role is also to engage in debates about technology regulation and issues such as data protection. But the world has moved on significantly since the appointment. The use of social media platforms in disinformation operations, the attempt by a social media platform to launch its own global currency and the debates about the involvement of technology companies with their governments and the setting of international industrial standards have all reinforced the idea of technology companies as geopolitical actors in their own right, with the privileges and responsibilities that such actors enjoy. This suggests that the remit of the Techambassador may have to expand to treat the tech sector as one would nation states, and not always as friendly ones. If technology companies refuse to cooperate fully in combating government-driven disinformation campaigns and fake news, they become complicit in those campaigns. If they fail to cooperate fully in combating crimes like human trafficking or child abuse, they become complicit in those crimes. And they risk being treated accordingly. But at the same time, the Techambassador should be actively seeking to recruit them into coalitions to shape the debates about Internet governance and regulation, as partners in the construction of international norms and rules of the game in cyberspace.

Similar questions arise about the technology and Internet companies themselves. If they are geopolitical actors, or are perceived as such, they need to develop their own techplomacy capacities. As shown by Facebook's clumsy launch of Libra/Diem, if you aspire to a role that others perceive as geopolitical, you need to develop a diplomatic approach or strategy, regardless of your self-perception as a commercial company. Likewise, technology companies navigating through the newly dangerous minefields of international industry standards meetings will need diplomacy as well as technical skills, not just to secure the right to set industrial standards in the first place, but also to survive the geopolitical disputes of the subsequent technology rollouts. Microsoft at least has recognised this and decided to appoint what is effectively a Microsoft ambassador to the EU. In a supreme irony it recruited Klynge himself as the first occupant of the position.

## 5.5 Conclusion

The rapidly developing digital technology, Internet and AI sectors are throwing up a host of political and geopolitical problems requiring diplomatic management at the international level. Major powers are increasingly turning technology and the technology industries into new battlefields for their geopolitical rivalry. The major technological and Internet companies increasingly appear like the oil companies of the twentieth century as geopolitical actors in their own right. Managing these three issues requires governments to develop an effective techplomacy to engage not only with each other but also with the technology and Internet companies themselves (these companies need to develop their techplomacy, as Facebook/Meta has learnt to its cost, although it is still not clear whether it has learnt the lesson as well as Microsoft).

The European Union and LAC countries are in a similar position as far as they do not have any global-level technology or Internet companies of their own. This potentially puts them at a disadvantage in dealing with the technology and Internet companies and limits their influence in key debates. This should provide an incentive for closer cooperation between the two continents on these issues. Denmark's appointment of a Techambassador was astute and far-sighted. It was not the publicity stunt some described it as at the time. Events have moved on, and the role and importance of the Techambassador and Techplomacy have increased. Internet and technology companies are no longer exclusively Western. Their geopolitical role is well-established. They can either become allies in dealing with the geopolitical and political challenges that new technologies pose, or they can become rivals to be dealt with accordingly. Cooperation between the EU and CELAC countries as ever is limited by the lack of effective regional institutions on the CELAC side to mirror the European Union. But the issues of techplomacy are going to increase rather than diminish and the EU, and even more CELAC countries, are at a marked disadvantage compared to countries like the USA and China. Despite the European Commission's best regulatory efforts, both risk becoming norm takers rather than norm makers. The potential of the Metaverse only makes the issue more urgent.

# References


Ball, M. (2022). *The Metaverse and how it will revolutionise everything*. Liveright.

Egloff, F. (2021). *Semi-state actors in cybersecurity*.

Lee, E., & Chau, T. (2017). *Telecom services: The geopolitics of 5G and IoT*. Jefferies Group.

Libicki, M. (2016). *Cyberspace in peace and war*. Naval Institute Press.

Miller, C. (2022). *Chip war: The fight for the World's Most critical technology*. Scribner.

Ministry of Foreign Affairs of Denmark. (2022). *How is Techplomacy operationalized. Office of Denmark's tech ambassador*. Retrieved May 12, 2022, from https://techamb.um.dk/the-techplomacy-approach

O'Hara, & Hall. (2021). *Four internets: Data, geopolitics and the governance of cyberspace*.

O'Neil, C. (2016). *Weapons of math destruction: How big data increases inequality and threatens democracy*. Crown.

Rid, T. (2020). *Active measures: The secret history of disinformation and political warfare*. Profile Books.

Riordan, S. (2018). The geopolitics of cyberspace: A diplomatic perspective. *Brill Research Perspectives in Diplomacy and Foreign Policy, 3*(3), 1–84.

Riordan, S., & Torres Jarrín, M. (2020). Techplomacy and the Tech Ambassador. *Global Policy Perspective Report*. European Institute of International Studies. Retrieved December 12, 2021, from https://www.ieeiweb.eu/wp-content/uploads/2020/01/Techplomacy-and-the-Tech-Ambassador.pdf

Singer, P. W., & Brooking, T. (2018). *Like war: The weaponization of social media*. Houghton, Mifflin, Harcourt.

Stephenson, N. (2011). *Snow crash*. Penguin.

Tapscott, D., & Tapscott, A. (2016). *Blockchain revolution: How the technology behind bitcoin is changing money*. Business and the World.

Zuboff, S. (2019). *The age of surveillance capitalism: The fight for a human future at the new frontier of power*. Profile Books.

# Chapter 6
# Science Diplomacy: Lessons from COVID-19 and Climate Change

**Abstract** The chapter will analyse the performance of science diplomacy in relation to the COVID-19 pandemic, and in particular how the pandemic highlighted the differences between science and policy. It will look at the lessons that can be learnt and at the challenges for science diplomacy in relation to COVID-19 going forward. It will then look at the lessons for science diplomacy that can be learnt from the experience of negotiating climate change agreements.

**Keywords** Science diplomacy · COVID-19 · Climate change

## 6.1 Science Diplomacy

The Austrian philosopher Wittgenstein famously commented that if a lion could talk, he would not be able to understand it (Wittgenstein, 1978). While not suggesting that diplomats and scientists cannot understand each other, they engage in the world in very different ways with very different criteria for knowledge and decision-making. In Wittgenstein's terms, they form part of different language games reflecting different "Lebensform" or life forms. Discussions and research into Science Diplomacy often overlook the diplomacy side in the enthusiasm for international scientific exchanges. But the COVID-19 pandemic emphasised the difference between science, policy and diplomacy. It demonstrated the extent to which science diplomacy functions along all three axes of engagement with diplomacy: agency, process and subject matter. Scientists were drawn into the roles of diplomats and policy advisors. Governments used science, and the science of COVID-19, as a tool to promote their broader diplomatic and foreign policy agendas. Even in the midst of the pandemic pursuing geopolitical rivalries often seemed more important than responding to the global health threat. Science itself had a good pandemic.

Policy-making and diplomacy had a less good pandemic. Science rapidly identified the genomic sequencing of the original COVID-19 variant and tracked subsequent variants in real time. Science successfully developed vaccines against and



M. Torres Jarrín, S. Riordan, *Science Diplomacy, Cyberdiplomacy and Techplomacy in EU-LAC Relations*, United Nations University Series on Regionalism 25, https://doi.org/10.1007/978-3-031-36868-4_6

treatments for COVID-19 in record time, which played a key role in converting a life-threatening pandemic into an endemic disease. If life in Western societies is at last returning to something like normality, science can claim a great deal of the credit. The problems arose when politicians leant on scientists to make policy for them, when governments sought to deal with COVID-19 at an international level and when science increasingly became an excuse for policymaking. When the decisions of politicians were criticised too often they claimed to be "just following the science". Not only was this a blatant attempt to divest themselves of any responsibility for the outcomes of the policies they had pursued, but it also dangerously conflated the two separate areas of science and policymaking. It undermined the distinction between science informing policy decisions and science being used as an excuse in policy debates (Gaston &Aspinall, 2020).

## 6.2 Science Diplomacy and COVID-19

In the second chapter we identified pandemic disease as one of the global issues in the New International Security Agenda (NISA). It has been identified as such since at least the beginning of the twenty-first century. Indeed, it has increasingly been identified as a hard security issue, which can threaten the security and stability of the state as well as the security and economic welfare of the individual in the state.

History supports this from stories of Mongols throwing the plague-infected corpses over the walls into the cities they were besieging (an early case of biological warfare) to the impact of AIDS/HIV on the deployment of African armies in peacekeeping operations in the 1990s (Riordan, 2008). An interesting example is the so-called Spanish Influenza, more technically H1N1 Influenza Virus A, and its impact on the end of WWI. It has long been recognised that it had a devastating impact on the world in 1918–1919 and must have had some impact on the end of the war. But more recent studies suggest that it may have had a more direct impact on military operations.

The St. Michael offensive in April 1918 was Germany's last attempt to win the war before the arrival of US soldiers. It came close to succeeding and was repelled in the end only at the second Battle of the Marne. The attack ran out of steam as the Germans ran out of combat-ready troops. A key factor appears to be the number of troops reporting sick or unfit for combat. Because German records did not specifically mention H1N1A as the cause of these "no shows" it is only recently that some military historians have suggested that H1N1A may have been a direct factor in Germany's failure to win the war in the first half of 1918 (Honigsbaum, 2020).

Despite this historical background and the widespread inclusion of pandemic disease in the New International Security Agenda, not all governments treated the outbreak of COVID-19 as a national security issue. Asian states, possibly because of their previous experience with SARS, did so. European states generally tended to deal with it as a public health issue. The distinction has important policy implications. Public Health policy is data driven and evidence based. Public Health policymakers expect to receive the data and make decisions based on the data. Their approach is cautious, with a bias towards inaction rather than action. National Security policymakers know that they do not have the luxury of waiting for the evidence. National Security decisions, like all foreign policy decisions, are taken in conditions of uncertainty with incomplete information (Cooper et al., 2008).

As Kissinger commented, the tragedy of foreign policy is that when you have greatest freedom of manoeuvre you have least information; when you have most information all of your options have been closed off by events. Asian governments did not need to wait for all the data. They responded with speed and agility to the events in China. European governments waited for the data that would allow them to take evidence-based public health policy decisions. They lost 2 months in the process. By the time the data arrived their policy options had been closed off by events and they were forced into more or less brutal lockdowns. The performance and outcomes of the Asian governments which treated COVID-19 as a national security issue were significantly better in the first phase of the pandemic than those of European governments which treated it as a public health crisis in the first phase of the pandemic.

There was considerable controversy in some countries, in particular within the UK, about the role of statistical modelling in driving policy on COVID-19. The decision to lockdown the UK economy in March 2020 was driven by models suggesting that over 300,000 people could die if the government took no action. This is problematic in several ways and could have serious implications for the use of mathematical and computer modelling in future science diplomacy, including in key areas like climate change. Successful computer modelling depends on two factors: the quantity and quality of the data and the design of the model. At this stage of the pandemic, data was both limited in quantity and of uncertain quality. With limited and unreliable testing, it was not clear how many people were contracting the disease, and what proportion of contagion was asymptomatic. This impacted on case fatality rates and thus the potential dangers of the virus. At the same time there were questions about the design of the models.

Modellers were confronted by a dilemma: use simplified models which did not attempt to build in all the nuances of behaviour or built ever more complicated models trying to capture every aspect of the contagion and human behaviour. This is a common dilemma for modellers. The former is more robust and more easily understood—the modellers understand the plumbing of the model. But it gives only a general feel for the evolution of the virus and does not produce forecasts sufficiently detailed to build policy decisions on. The latter appears to give more detailed forecasts, but models quickly become so complicated that the modellers do not understand their own models, or the implications changes in one part of the model have elsewhere. They also assume a degree of linearity in human societies which many analysts regard as complex autonomous systems and thus essentially non-linear (Gardini, 2020).

It is not clear whether the modellers or governments' scientific advisors made clear the difficulties with these models and the reasons for questioning their output when presenting these model-based forecasts to government ministers. History would suggest not. Similar modelling techniques are widespread in the financial industry. They are used to assess risk in financial markets and thus help formulate the investment policies of banks and investment funds. Most of these models were based on the assumption that risk in financial markets follows a normal distribution whereas the evidence suggests it follows a power law distribution, but a power law distribution does not allow for predictions which is what the models are for. Most of the time this does not matter but when there is volatility in the markets the models tend to blow up. The executives taking the key investment decisions did not know this (and may not have understood if it had been explained to them). But they were encouraged to rely on the models when taking their decisions (O'Neil, 2016).

Government ministers at the beginning of the pandemic may have been in a similar situation. They were taking decisions on unreliable forecasts from models using poor data and overcomplicated model design. This matters in the wider context of science diplomacy because it may undermine confidence in mathematical computer modelling as a whole. Given that much of the evidence in policy debates about climate change also depends on computer modelling this could complicate the task of those science diplomats seeking to negotiate international agreement on carbon emissions targets and other international climate change agreements.

One of the features of scientists converted into policy advisors, especially prominent in the UK but also seen elsewhere, was intolerance of dissent once a policy was established. This intolerance appeared on both sides of the debate, among those criticising the policies as much as those advocating them. Given that governments frequently abandoned their own policymaking responsibilities to "follow the science", scientists were transformed from scientists to policy advisors and in effect to policymakers. This meant that policy was driven only by concerns about fighting the pandemic and reducing contagion levels without taking full account of wider economic, social or psychological, or even geopolitical, factors. Severe economic damage was committed in Western countries in the name of combatting COVID-19. Other health issues, such as the diagnosis and treatment of cancer were left to one side (Tooze, 2021). This is not how Western governments have reacted to previous pandemics.

The Asian Flu (H2N2) in 1957 and the Hong Kong Flu (H3N2) in 1968 both killed well over a million worldwide (out of a much smaller global population). But in neither case did Western governments take drastic action or consider closing down entire economies (possibly because of the Cold War and also citizenries who expected less of their governments). This points to a danger in science diplomacy (and other specialist diplomacies): scientists advocating certain policies, e.g. relating to climate change, understandably focus on the issue that concerns them without taking account of the wider economic or geopolitical context.

The "diplomacy" of science diplomacy must take account of this wider context if it wants success in its advocacy. Activists will not be happy with the outcomes but without taking account of the wider context there will be no outcome.

International and regional collaboration in the face of the virus was patchy at best. Western governments which had not used the time which elapsed between the outbreak of the virus in Wuhan and its arrival on their shores to build up stocks of testing kits and protective clothing fought each other to secure whatever stocks were available. There were unedifying reports of diplomats trying to outbid each other on airport runways to secure equipment.

Countries in the global south stood little chance of securing testing or protective equipment. Governments sought scapegoats to distract from their own failings. Western governments criticised China for trying to conceal the outbreak suggesting that rather than being naturally occurring the virus had leaked from a medical research centre in Wuhan. China's so-called "wolf warrior diplomats" hit back criticising the performance of Western governments (Gaston & Aspinall, 2020).

The World Health Organisation (WHO) was another target of criticism with the USA withdrawing from the organisation in July 2020 (it returned in January 2021 once President Biden was inaugurated as US President). Even within the EU, there was little collaboration at the European level in the first phase of the pandemic with individual member states adopting whatever measure they thought necessary, including closing borders within the Schengen area, with the European Commission reduced to rubber stamping them after the event. Interestingly, citizens looked to their national governments rather than the European Union to protect them in a crisis. As European nations struggled individually to contain the virus there was limited if any scope for collaboration between the EU and LAC countries, some of which would witness the most appalling situations (Farrar & Ahuja, 2021).

Some countries sought to take advantage of the lack of international collaboration to gain geopolitical influence. Thus, when Italy was suffering the worst of the pandemic both Russia and China supplied masks and other protective equipment in high-profile operations in which the national origins of the equipment were prominent. Similar operations were carried out in other European countries.

In Spain, the Chinese community made a point of delivering masks to hospitals. There was the irony of countries like Russia and China providing Western European countries with emergency aid. There was also a sense in which China was seeking to improve its image after the criticism that it had tried to conceal the virus in the first place. This provision of protective equipment evolved into what was called "vaccine diplomacy". As noted above, the development of a range of vaccines against COVID-19 and its variants in record time was a major triumph for science. But the policy and diplomacy sides were more controversial.

Western governments understandably gave priority to vaccinating their own citizens over providing vaccines for other countries. But the demand for absolute security against COVID-19 and the perception that mass vaccination was the only path to a return to economic normality meant that Western countries prioritised multiple vaccinations of their citizens before providing vaccines to other countries, especially those in the global south. European countries insisted on booster injections for their citizens despite the WHO arguing for instead providing vaccine doses for countries in greater need. Once again China and Russia stepped in offering to ship large quantities of their home-produced vaccines to countries in the global south, including in Latin America. While there were some concerns about the efficacy of the Chinese vaccines, the Russian Sputnik vaccine was praised in the pages of the leading medical journal *The Lancet*.

The Chinese and Russian governments were in effect willing to sacrifice the well-being of their own citizens to gain influence with other governments and show up the lack of solidarity shown by Western governments. The Chinese did so believing that they had successfully controlled COVID-19 at home through their "Zero COVID" approach. They were to pay for this subsequently when later variants of COVID spread rapidly through a population with low levels of vaccination. Western governments complained that this was a cynical effort to take advantage of the pandemic for geopolitical ends, but the recipients did not complain. Once again Western governments, apparently driven by only science, failed to incorporate wider geopolitical factors into their policymaking. It is not clear to what extent vaccine diplomacy was effective, or the extent to which countries which received doses of the Sputnik vaccine have been more inclined to take a neutral stand on Russia's invasion of Ukraine.

The previous chapter explored techplomacy: how governments and their diplomats can engage with digital technology and Internet giants. The COVID-19 pandemic drew attention to the fact that the problem did not lie just with digital technology companies. Pharmaceuticals and biogenetics are also high technology whose companies often wield a power and influence with which many governments cannot compete. The scope of techplomacy may therefore need to be expanded to include a subset of "pharmaplomacy"—the ways in which governments and their diplomats engage with pharmaceutical companies (and the ways in which pharmaceutical companies engage with governments).

The history of the COVID-19 pandemic certainly showed a lack of such "pharmaplomacy" on both sides. We have already mentioned the conflict between the European Commission and Astra Zeneca, in which the Commission accused Astra Zeneca of failing to meet its contractual commitments. There were lesser conflicts between the Commission and other pharmaceutical companies. In the case of Astra Zeneca, these disputes and the consequent comments by European politicians, undermined confidence in the vaccine and reduced European vaccine coverage. There will be more pandemics and other international medical emergencies.

Pharmaceutical companies will always be a key element in governments' responses. But pharmaceutical companies operate at a global rather than national level and often prioritise profit rather than perceived medical necessity. Governments and their diplomats will need to engage with them on a global and often multilateral level, convincing them to align their priorities with those of the governments (Ferguson, 2021).

The COVID-19 virus was challenging for multilateral diplomacy in two senses: its impact on the way in which diplomacy operates and in demonstrating the weakness of multilateral institutions. As already pointed out, the WHO came under heavy criticism and not just from the Trump administration (which accused it of being controlled by China). Much of the criticism did not understand how multilateral institutions operate in the UN system, and the limits on the extent to which the institutions' employees (the secretariat) can criticise individual member states. But the WHO did demonstrate itself overly bureaucratic and slow to react to a fast-moving crisis. The strict rules established following the SARS outbreak about designating a pandemic seemed to slow down the formal designation in ways that undermined the international response (and provided an excuse for Western governments for their own slow responses).

The WHO also demonstrated its weakness when its advice was ignored, for example on supplying vaccine doses to developing countries rather than offering booster vaccinations, demonstrating that its role was essentially coordination and exhortation. Other multilateral institutions similarly showed themselves slow to respond.

The European Commission insisted on following pre-existing rules on procurement which delayed the purchase of vaccination doses and then got into an unseemly row with Astra Zeneca about the provision of vaccines. Critical comments about Astra Zeneca by some European leaders may have slowed the uptake of vaccinations in some European countries. LAC countries were pretty much left to their fate.

Meanwhile, the COVID-19 virus had a significant impact on how diplomacy, and in particular multilateral diplomacy, is carried out. COVID-19-related lockdowns and travel restrictions meant that many meetings had to be moved online. European Councils, the UN Security Council and even the UN General Assembly were held online. The face-to-face contact which is central to diplomacy was lost. Diplomats could no longer have the discreet meetings in the margins of meetings (the coffee shop huddle), which drive forward many negotiations. Diplomats complained of the impossibility of networking online. There were some advantages. Online meetings were cheaper and more flexible. People did not need to be physically assembled in a meeting room. International travel was unnecessary. Participants from different countries could meet online. Diplomats who had always maintained that doing diplomacy online was impossible were forced to break that taboo and discovered that it was possible to diplomacy through online platforms. With the taboo broken, more innovative diplomats looked for online alternatives to face-to-face diplomacy.

WhatsApp and Snapchat replaced the discreet conversations in the margins. Some speculated about the parallels between diplomatic networking and online dating applications: if people can find partners online surely diplomats can find relevant contacts. The general consensus among diplomats is that diplomacy, and thus science diplomacy, will not return completely to pre-pandemic practices but nor will it remain completely online.

A hybrid diplomacy will develop which combines more online meetings with the return of face-to-face diplomacy. This will not be without problems. Concerns will remain about the reliability of off-the-shelf software for online meetings and how security can be guaranteed. Concerns also focus on internet infrastructure and the reliability of Internet connections, particularly in the global south. If more diplomatic negotiations are to be conducted online, those without reliable internet access will be prejudiced and their potential influence over decisions affecting their future undermined. This is likely to be a greater concern among LAC countries than within the EU.

The interaction of COVID-19 with diplomacy and science diplomacy does not relate only to the past. Future pandemics can be anticipated (pandemics and potential pandemics since 1990 include HIV/AIDS, SARS, MERS and Bird Flu).

Concerns are being raised at the time of writing about the spread of Monkey Pox, although no one is suggesting yet that this could be a potential pandemic. Nevertheless, another pandemic will emerge at some point, probably driven by the increasing human invasion of animal habitats. It has been pointed out for some time that modern lifestyles contribute to the spread of infectious diseases. In particular, cheap air travel and mass tourism accelerate the spread of diseases.

In the nineteenth century, the weeks needed for travelling by ship meant that symptoms would emerge during the voyage allowing diagnosis while at sea and isolation on arrival (in the quarantine docks). The mere hours needed for air travel means that an infected person can arrive home before the symptoms emerge. They are mixing with friends and relatives as the symptoms do emerge and probably treat them with a paracetamol in the first instance. When they do go to a local doctor, he may well not diagnose the disease. By the time they die, and an autopsy identifies the disease it may already be widely spread in the community. Mass tourism also changes the calculation.

Thirty years ago, mass tourism remained regional, with the vast majority of Europeans taking their holidays within Europe. Increasingly tourism is global rather than limited to the West and tourists seek to visit ever more exotic places where diseases against which they have no prior immunity are more likely. During the pandemic extensive health security measures were introduced into airports. They could be chaotic and differ from airport to airport. As the pandemic has receded, they have become even more chaotic with different rules about masks, for example in different airlines operating from the same airport. There is likely to be a discussion about whether permanent health security measures should be introduced into airports analogous to those aimed at preventing terrorism. There will be negotiations to ensure that any measures introduced are homogenous across all regions of the world. Such negotiations need to be informed by the science to ensure that such measures are effective and needless or ineffective measures are not introduced: a task for science diplomacy.

There are other tasks for science diplomacy in the wake of the COVID-19 Pandemic. Chapter 4 explored the impact on science diplomacy of the conflict between the New International Security Agenda and the more traditional Geopolitical Agenda. The same conflict has undermined efforts to learn the lessons from COVID-19. Exploration of the phylogenesis of the virus has been bedevilled by the geopolitical blame game with Western politicians seeking to divert attention from their own failings by implying that the virus escaped the Wuhan research facility. The tensions generated make it unlikely that the origins of the disease will ever be clearly established. Nevertheless, public opinion is likely to demand international rules about the management and security of such medical research facilities.

Science diplomacy will need to find ways of avoiding such discussions being infected by geopolitical rivalry. This forms part of a broader challenge for science diplomacy: to create geopolitically free spaces where scientists can come together to discuss issues like the origins and spread and of pandemic diseases, thus providing agreed science on the basis of which policy discussions can take place. These policy spaces should not be limited to scientists from Western countries. Indeed, this could form a profitable area of cooperation between the EU and LAC countries into which scientists from other regions could be integrated. Key to the success of this project, and a constant theme of this book, is the need to keep clear the distinction between science and policy.

In confronting the challenge posed by COVID-19 and other global challenges in the New International Security Agenda, science diplomacy can benefit from the lessons learnt from the climate change debate. As noted in Chap. 3, concerns about climate change and global warming first circulated on networks constructed to promote international scientific exchanges. At this stage, there were exchanges between scientists rather than science diplomacy as such. These exchanges built up a solid body of scientific knowledge and understanding that garnered more credibility with policymakers than lobbying by activists would have done. It also made it harder for corporate interests, especially in the oil industry, to undermine the scientific case for climate change and global warming (although they still tried). But this is related to establishing the scientific case for climate change, not the policy debate on what to do about it. The exchanges on the scientific networks generated the science that would inform the policy debate but could not decide it. The policy debate on what to do about climate change was no longer just about science, or even principally about science. It had also to take into account a broad range of economic, social, technological and geopolitical factors. Scientists taking part in the policy debates were no longer doing so as scientists but as policy advocates. Many scientists and activists were appalled by the extent to which including these other factors diluted the scientific debate about climate change, but without taking them into account there would have been no agreement. In this lies the difference between science and science diplomacy. The Paris Accords on climate change were ultimately made possible by a geopolitical deal between Presidents Obama and Xi Jinping.

The diplomatic processes of the climate change talks also have lessons for wider science diplomacy. The talks included not only government diplomats but also scientists, corporate executives, NGOs and citizen activists. They required effective multistakeholder diplomacy which involved engaging with the full range of state and non-state actors (Riordan, 2003). Diplomats had to network with the different actors attempting to construct heterogenous and multi-level coalitions in support of favoured outcomes. Such coalitions could not be based on shared values or ideologies. It was extremely unlikely that a successful coalition could be constructed between state actors, corporations, NGOs and citizen activists in which each of the partners shared the same values. But all of these different elements had to be included if the policy was to be advanced. Coalitions, therefore, had to be constructed on the basis of shared preferred outcomes. This reflects an older diplomatic tradition based on interests.

As Lord Palmerston said in the middle of the nineteenth century "Therefore I say that it is a narrow policy to suppose that this country or that is to be marked out as the eternal ally or the perpetual enemy of England. We have no eternal allies, and we have no perpetual enemies. Our interests are eternal and perpetual, and those interests it is our duty to follow". This is often portrayed as unacceptably cynical. But a foreign policy based on shared values or ideologies limits agreement to those countries with whom one shares values. This makes global agreements all but impossible, which becomes crucial when dealing with the global issues of the New International Security Agenda. Part of the art of diplomacy lies in trying to identify how other nations see their interests (and how they perceive your interests), identifying where those interests overlap and, on that basis, constructing agreements. In the case of science diplomacy, the art lies in identifying shared preferred outcomes and building coalitions around them. Inevitably no one is completely happy with the final outcome. But one of the tragedies of foreign policy is that it is often reduced to identifying and securing the best out of a series of bad outcomes. These lessons could lie at the heart of a joint EU/LAC science diplomacy.

## References

Cooper, A., et al. (Eds.). (2008). *Global governance and diplomacy: Worlds apart?* Palgrave Macmillan.

Farrar, J., & Ahuja, A. (2021). *Spike: The virus vs the people*. Profile Books.

Ferguson, N. (2021). *Doom: The politics of catastrophe*. Penguin.

Gardini, G. L. (2020). *The world before and after Covid-19: Intellectual reflections on politics*. Diplomacy and International Relations, European Institute of International Studies.

Gaston, S., & Aspinall, E. (2020). *Coming together and pulling apart: How the coronavirus pandemic tested global institutions and reshaped geopolitical realities*. The British Foreign Policy Group.

Honigsbaum, M. (2020). *The pandemic century: A history of contagion from the Spanish Flu to COVID-19*. Penguin.

O'Neil, C. (2016). *Weapons of math destruction: How big data increases inequality and threatens democracy*. Crown.

Riordan, S. (2003). *The new diplomacy*. Polity.

Riordan, S. (2008). The new international security agenda and the practice of diplomacy. In A. Cooper et al. (Ed.), *Global governance and diplomacy: Worlds apart?* Palgrave Macmillan.

Tooze, A. (2021). *Shutdown: How COVID shook the world's economy*. Allen Lane.

Wittgenstein, L. (1978). *Philosophical investigations*. Blackwell.

# Chapter 7
# Science, Technology and Innovation in the EU-LAC Cooperation and the Agenda 2030: The Sustainable Development Goals

**Abstract** The chapter looks at science diplomacy in relation to the SDGs and Agenda 2030, it also interrelates the EU-CELAC inter-regionalism focused on achieving the objectives set out by the United Nations. How multistakeholder diplomacy functions in relation to the SDGs and Agenda 2030 will be a key focus of the chapter and how EU-CELAC cooperation can reinforce this and the commitment of the two regions to achieving them.

**Keywords** SDGs · Agenda 2030 · EEAS · EU-CELAC interregionalism · Science · Technology · Innovation · EU science diplomacy · EU international cooperation

## 7.1 EU Science, Technology and Innovation Practices and the Sustainable Development Goals

History shows us that international science cooperation is a positive way to build bridges between countries and contributes to regional integration. The Higgs particle was discovered at the European Organization for Nuclear Research (CERN), an organisation formed to build the foundations for European science after the Second World War by bringing together former adversaries. Along with CERN, the Abdus Salam International Centre for Theoretical Physics (ICTP) is one of the oldest international research institutions (Quevedo, 2013).

The European External Action Service closely works with the European Commission in order to enhance the science diplomacy dimension. Article 180 of the Treaty on the Functioning of the European Union (TFEU) includes the implementation of research, technological development and demonstration programmes, by promoting cooperation with and between undertakings, research centres and universities as well as the promotion of cooperation in the field of European Union research, technological development and demonstration with third countries and international organisations (TFEU Article 180).



M. Torres Jarrín, S. Riordan, *Science Diplomacy, Cyberdiplomacy and Techplomacy in EU-LAC Relations*, United Nations University Series on Regionalism 25, https://doi.org/10.1007/978-3-031-36868-4_7

In 2021, the European Commission announced the EU's strategic approach to enhancing EU international cooperation in research and innovation, underlining the significance of international cooperation and the role of the EU that derives from being a global leader in research and innovation (European Parliament, 2019).

According to European Commission the EU is a global leader in research, development and innovation, accounting for 20% of world expenditure on research, 32% of high-impact publications and 25% of patent applications (European Commission, 2018). The EU is the third biggest Research and Development and innovation investor in the world, after the USA and China.

According to the Global Innovation Index 2021 rankings there are 8 EU member states in the Top 20. The ranking also classifies as the Top three innovation economies by region Switzerland, Sweden and the UK in the European region and Chile, Mexico and Costa Rica in the Latin America and the Caribbean region (World Intellectual Property Organization, 2021).

Horizon 2020 (H2020) is a European programme that focuses on research and innovation cooperation. During the period 2014–2020 the programme was worth about 80 billion euros. Horizon 2020 responds to strategic objectives such as using international science cooperation as an instrument of soft power and a mechanism for improving relations with key countries and regions (European Commission, 2012). H2020 is the financial instrument implementing the Innovation Union, a Europe 2020 initiative aimed at securing EU's global competitiveness. H2020 derived from the commitments made under the Innovation Union, the European Research Area Framework and the recommendations of the Seven Framework Program (FP7). The Innovation Union introduced a strategic and broad approach to innovation by including actions that aimed to tackle both the supply and demand side elements of the innovation ecosystem: public sector, businesses, academia and finance (European Commission, 2015).

The European Commission aimed to strengthen the EU's global position in research, innovation and technology through proactive international cooperation and signing Science and Technology agreements with some more strategic partners such as the USA, Australia, Canada, China, Korea, India, New Zealand, Russia and South Africa. In the case of Latin American countries, the EU signed agreements with Brazil, Chile and Mexico. At the same time, the European Commission opened policy dialogue on Science, Technology and Innovation at the regional level with Africa, ASEAN, Central Asia, Gulf, Latin America and the Caribbean and Pacific (European Parliament, 2019).

For the new period 2021–2027, the EU has created Horizon Europe which replaces H2020 and has a budget of 95.5 billion euros and is divided into four pillars and 15 components that will support all the areas of research and innovation: excellent science, global challenges and industrial competitiveness, innovative Europe and widening participation and strengthening the European Research Area (European Commission, 2021a). This budget also includes 5.4 billion euros from the Next Generation Fund to support the green and digital recovery from the COVID-19 crisis.

Under the Horizon 2020 84% of investments address Sustainable Development Goals (SDGs) (European Commission, 2021b). European third countries with capacity in science, technology and innovation can participate in the programme. Latin American and Caribbean scientists can take part in the Horizon Europe programme. They may also apply for Marie Sklodowska-Curie Actions and for the European Research Council grants which are for the most prestigious and highest quality researchers. In the first work programme (2021–2022), LAC countries can present projects in the following areas (European Union, 2021):

- Public policies and indicators for well-being and sustainable development.
- New generation of full-electric urban and peri-urban Bus Rapid Transit systems to strengthen climate-friendly mass transport (2Zero).
- Stimulating Road Transport research and innovation dissemination and implementation in Europe and around the World.
- Quantify impacts of the trade in raw and processed biomass on ecosystems, for offering new leverage points for biodiversity conservation, along supply chains, to reduce leakage effects.
- Understanding the impacts of and the opportunities offered by digital transformation, new emerging technologies and social innovation with respect to biodiversity.
- Network for nature: multistakeholder dialogue platform to promote nature-based solutions.
- Assessing the nexus of extraction, production, consumption, trade and behaviour patterns and of climate change action on biodiversity in the context of transformative change.
- Assessing the socio-politics of nature-based solutions for more inclusive and resilient communities.
- Developing nature-based therapy for health and well-being.

As we can observe all these calls intend to comply with the Sustainable Development Goals.

In the H2020 framework two consortia were financed with a focus on science diplomacy: Inventing a shared Science Diplomacy for Europe (InsSciDe) and Using Science for/in Diplomacy for Addressing Global Challenges (S4D4C).

InsSciDe is an interdisciplinary consortium centred on the development of a shared science diplomacy across Europe through international and interdisciplinary research. InSciDer considers that science diplomacy is at the core of work to achieve the UN's Sustainable Development Goals through a multi-inter-disciplinary understanding of science itself, which takes into account a diversity of knowledge systems within academia and society (InsSciDe, 2019).

S4D4C is a project that supports science diplomacy as a means to foster EU foreign policy goals and EU commitments with the global challenges (Meyer, 2021) as the SDGs. In 2019, S4D4C promoted the Madrid Declaration on Science Diplomacy which proclaims a common vision of the future of science diplomacy and emphasises the benefits science diplomacy can bring to tackling the global challenges. The Madrid Declaration has among its objectives to foster agreement and raise awareness about the need to strengthen science diplomacy strategies and practices worldwide for the support of universal scientific and democratic values (S4D4C, 2019).

In 2021, S4D4C, InsSciDE and EL-CSID launched the European Union Science Diplomacy Alliance with the support of several founding members. The Alliance emerged as a result of the S4D4C policy brief "Nurturing the EU Science Diplomacy Community: The launch of an EU Science Diplomacy Alliance for Addressing Global Change". The policy brief identifies different types of science diplomacy stakeholders—governmental, intergovernmental and supranational, research and academic, from the private sector and civil society—and maps the current science diplomacy networks at the subnational, national, sub-global and global levels (Elorza et al., 2021).

**Science Diplomacy Subnational Networks**
- Spain: Scitech Diplohub-Barcelona
- United States: Science and Technology Diplomatic Circle (S&TDC) Boston

**Science Diplomacy National Networks**
- United Kingdom: UK Science and Innovation Network
- Switzerland: Swissnex network
- Germany: German Academic International Network (GAIN)
- Malaysia: Scientific Malaysian
- Spain: Red de Asociaciones de Investigadores y Científicos Españoles en el Exterior (RAICEX)
- Greece: knowledge and Partnership Bridges (Gefyres Gnosis Kai Synergasias) Initiative

**Science Diplomacy Sub-global Networks**
- United States: Society for the Advancement of Science and Technology in the Arab World (SASTA)
- United States (funding by European Commission): European Diasporas Network in the USA
- Finland: The University of Artic (UA)
- Spain (funding by Ibero-American countries): Programa Iberoamericano de Ciencia y Tecnología para el Desarrollo (CYTED)

**Science Diplomacy Global Networks**
- Switzerland: The Intergovernmental Panel on Climate Change (IPCC)
- Germany (funding by UNESCO): Intergovernmental Science-Policy Platform for Biodiversity and Ecosystem Services (IPBES)
- New Zealand: Foreign Ministries S&T Advice Network
- United States: Networks of Diasporas in Engineering and Science (NODES)
- United States: Science Diplomats Club of Washington DC
- Austria: Big Research Infrastructure for Diplomacy and Global Engagement through Science (Bridges)

## 7.2 LAC Science, Technology and Innovation Practices and the Sustainable Development Goals (SDGs)

From the Economic Commission for Latin America and the Caribbean (ECLAC), we found several initiatives and projects related to promoting good practices in the areas linked to science, technology and innovation and Sustainable Development Goals (SDGs). For example, between 19 and 21 April 2021, ECLAC held the "Training workshop for Latin America on policies and policy instruments in Science, Technology and Innovation for SDGS". The aims are to raise awareness and understanding of key aspects of STI policies as well as innovation and how they relate to the SDGs. The workshop is aimed at the professionals and policymakers from the public sector with focus on science, technology and innovation policies (United Nations, 2021a, b).

In the framework of the 36th Session of the ECLAC, in 2016, its Member States adopted (ECLAC Resolution 700, 2016) which established the Forum of the Countries of Latin America and the Caribbean on Sustainable Development as a regional mechanism to implement the 2030 Agenda for Sustainable Development, including the Sustainable Development Goals, as well as the Addis Ababa Action Agenda. Since the creation of the Forum a fourth edition has been organised. The last was the Fifth Meeting in San Jose on 7–9 March 2022. The forum seeks to provide useful peer-learning opportunities through exchange of good practices and discussion of common goals. Usually, the forum conclusions are submitted directly to United Nations High-Level Political Forum on Sustainable Development which is a mechanism that meets every July at the United Nations headquarters in New York.

In February 2022 ECLAC presented the Fifth report on regional progress and challenges in relation to the 2030 Agenda for Sustainable Development in Latin America and the Caribbean (ECLAC, 2022). This report presents specific proposals in four areas:

1. Strengthening multilateralism, particularly in the area of financing for development.
2. Improving implementation of productive, social and environmental policies of national and regional scope.
3. Building up the resilience of institutions.
4. Overcoming conflicts through agreements and compacts.

An important point to note is that the LAC is a region committed to promoting and defending sustainable development both regionally and globally. In Escazú, Costa Rica the Regional Agreement on Access to Information, Public Participation and Justice in Environmental Matters in Latin America and the Caribbean 2018 was adopted. Among its aims is to guarantee an effective implementation in LAC of the rights of access to environmental information, public participation in the environmental decision-making process and access to justice in environmental matters, as well as in the creation and strengthening of capacities and cooperation, contributing to the protection of the right of every person of present and future generations.

The process of the adoption of the agreement was initiated at the United Nations Conference on Sustainable Development (Rio+20) that adopted the (Declaration of the application of Principle 10 of the Rio Declaration on Environment and Development 2012) signed by Chile, Costa Rica, the Dominican Republic, Jamaica, Mexico, Panama, Paraguay, Peru and Uruguay. Between 2014 and 2018, a Negotiating Committee was established composed of 24 countries of the region with the participation of the public. The Committee adopted its Organization and Work Plan in 2015 and the Modalities of Participation of the Public in 2016.

Gual Soler (2020) indicates six Latin American initiatives linked with science, technology and innovation practices, three at regional level and three in the university sector and in the diplomatic academies sphere:

**Regional Initiatives**

1. In 1984, Ibero-American Program of Science and Technology for Development (CYTED) was created by the Interinstitutional Framework Agreement signed in Madrid by the Ibero-American Community of Nations. Since 1995 the programme has been formally included among the Cooperation programmes dependent on the Ibero-American Community of Nations Summits (Cyted, 2018). CYTED promotes cooperation in science, technology and innovation, its objectives are to foster common projects between private sector and academia. The results of the programme include the generation of projects on Research+Development involving companies, universities, research centres and experts from the CYTED cooperation platform who have access to international funds. Since 2018, the science diplomacy is part of the agenda in the Ministerial Meetings and Senior Officials of Science, Technology and Innovation in the Ibero-American Summits.
2. In 1992, American countries held a meeting in Montevideo and signed the Montevideo Declaration which created the Inter-American Institute for Global Change Research. Twelve American governments requested the Institute to coordinate international scientific and economic research on consequences of global change in the Americas (Montevideo Declaration, 1992).
3. In 2016, the first edition of the Latin American and Caribbean Open Science Forum was held in Montevideo. The forum is a regional space for debate and exchange and is the result of the collaboration of regional organisations such as Inter-American Development Bank, Ibero-American Secretariat, Organization of Ibero-American States for Education, Science and Culture, Iberoamerican Network of Science and Technology Indicators, AUGM and CUIB with the executive secretariat of UNESCO. Since foundation the forum has organised three editions in 2016, 2018 and 2021.

**Universities and Diplomatic Academies**

4. In 2018 and 2019, Mexico, Argentina, Chile and Panama created capacity-building programmes and courses on science diplomacy.
5. In 2019, the University of Sao Paulo (Brazil) created the Sao Paulo School on Science and Innovation Diplomacy.

6. In 2020, Externado University of Colombia included a science diplomacy module in its New Diplomacies course.
7. In 2020, the National Autonomous University of Mexico implemented the Bachelor in International Relations a Science Diplomacy course.

It is important to note that CYTED was the first EU-LAC interregional dialogue mechanism with a focus on science and technology but it is true that it only included Spain and Portugal from the European side. This mechanism also created a new model for South-South cooperation between Spain, Portugal and Latin American countries (Sebastián, 1992). Since the foundation had the institutional support of very important international organisations such as the United Nations Economic Commission for Latin America and the Caribbean (ECLAC), the Organization of American States (OAS) and UNESCO these three organisations participated as observers (United Nations, 1984). Initially, the programme was coordinated in Spain and received support and funding from the Ministry of Education and Science and, in 1992 the Ibero-American Conference of Head of State and Government resolution ratified CYTED as an instrument for regional integration.

During the last two decades, LAC countries had positive economic, social and environmental levels. But the pandemic caused by COVID-19 has modified all these indicators. Every time there is an economic crisis, countries usually adjust their national budgets, reducing their spending and investment. But the current crisis has served to change the development model. The countries must take advantage of investing in science, technology and innovation, favouring changes in design, production and business models, betting on a circular economy and a green economy.

Since the SDGs adoption in 2015 most countries in the LAC region have shown progress in the fulfillment of the SDGs, although still insufficient if they want to achieve the SDGs by 2030. According to the SDG Index for Latin America and the Caribbean the countries improved by 0.67 points on the 2019 SDG Index representing an advance of 1.09 percentage points from the 2015 baseline, but if the countries maintain this rate the 17 SDGs in the 24 countries of the region would take more than 50 years (SDG Center for Latin America and the Caribbean, 2020).

Science, technology and innovation issues are relatively new in LAC policies. The incorporation of these issues is mainly due to the bi-regional dialogue EU-LAC summits and after EU-CELAC summits. Of the 33 CELAC Member States, only 5 countries have ministries of science, technology and innovation: Argentina, Brazil, Costa Rica, Cuba and Venezuela. The rest of the countries have national councils or secretariats on science and technology or technology and environmental and science, technology and innovation (RICYT, 2022).

## 7.3 EU-CELAC Relations and 2030 Agenda for Sustainable Development

The EU has recognised its responsibility to promote the 2030 Agenda for Sustainable Development regionally and globally (Council of the European Union, 2017). Over recent decades the EU and LAC countries have worked for a just and more balanced global governance. The cooperation between EU and CELAC has been committed to a sustainable world, promoting the transformation towards a green economy, transition to a circular economy, blue economy, sustainable and smart cities, decent work and competitive responsible businesses and investing in knowledge, innovation and human capital (European Parliament and Council of European Union, 2019).

Among the 17 SDGs poverty eradication and addressing inequalities are central themes in the 2030 Agenda which are key priorities for Latin America and Caribbean countries. The pandemic caused by COVID-19 has generated serious short- and long-term effects on supply and demand at the aggregate and sectoral levels (ECLAC, 2020a). The region is also characterised by weak and fragmented health and social protection systems and growing marginalised urban settlements that lack access to basic services (ECLAC, 2021).

The first EU-CELAC Summit, Santiago de Chile in 2013, was held under the theme "Alliance for Sustainable Development: Promoting Investments of Social Environmental Quality". Both regions expressed their commitment to achieving sustainable development in three dimensions: economic, social and environmental, as well as reaffirming the role of the UN General Assembly, ECOSOC and the UN Environment Programme, the prompt establishment of the High-Level Political Forum on Sustainable Development and the UN Development Agenda beyond 2015 that included the implementation of the Rio+20 mandate for the elaboration of a set of Sustainable Development Goals (Santiago Declaration, 2013).

In 2015, the EU-CELAC countries reiterated their adherence to the United Nations Framework Convention on Climate Change and its principles, highlighted the adoption of the "Lima Call for Climate" during COP20/CMP10 and reaffirmed the importance of the bi-regional strategic partnership and cooperation to adopt policies that promote trade and investment between CELAC and EU countries convinced that this will contribute to ensuring sustainable development (Brussels Declaration, 2015). Another important point was the introduction of the "New Innovation Pillar" in the bi-regional agenda.

In 2016, the EU launched its Global Strategy which indicated a revival of the Atlantic Space with the participation of Latin America and the Caribbean countries (EEAS, 2016) together with the EU and the USA. This could have meant an increase in the political weight of LAC within the traditional transatlantic relations and ultimately as an actor in the international arena. However, it did not happen due to two external factors: the Brexit process and the Administration changes in the USA firstly, with President Obama and then with President Trump.

Under Mexico's Pro-Tempore Presidency of CELAC, the Senior Officials Meeting of the EU-CELAC Joint Initiative on Research and Innovation (SOM EU-CELAC JIRI) was held on 30 October 2020. The meeting defined the direction of future Research and Innovation cooperation between the two regions outlining priority areas in the four pillars of mobility of researchers, cooperation of research infrastructures, global challenges and innovation.

The European Commission approved a decision on the EU-CELAC 2021–2023 Strategic Roadmap for the implementation of the Brussels Declaration and EU-CELAC Action Plan on Science, Technology and Innovation. The countries agreed that innovation is an essential ingredient for long-term prosperity of citizens and for the promotion of sustainable and inclusive development. In the framework of the EU-CELAC Summits the EULAC Innovation Cooperation Conference was organised in Brussels whose recommendations were to build enhanced innovation cooperation and make better use of existing initiatives, such as mobility of researchers, technicians and innovators between both regions and support the transition to a more productive, diversified and environmentally sustainable economy (European Commission, 2021c).

## 7.4 How Multistakeholder Diplomacy Functions in Relation to the SDGs and Agenda 2030

In 2015, after eight rounds of intergovernmental negotiations and thanks to an active participation of stakeholders the 2030 Agenda for Sustainable Development was approved, and the 17 Sustainable Development Goals (SDGs) were launched to replace the Millennium Development Goals (MDGs). Both represented a new development model capable of eradicating extreme poverty, generating quality employment, ensuring healthy lives and promoting well-being for all at all ages, and tackling the climate crisis (ECLAC, 2020b).

According to the General Assembly resolution 70/1 on the 2030 Agenda for Sustainable Development, the President of the United Nations Economic and Social Council (ECOSOC) will convene the meetings of the Multi-Stakeholder Forum on Science, Technology and Innovation for the SDGs (STI Forum) once a year to discuss the themes mentioned above. This resolution mentioned that all countries and all stakeholders acting in collaborative partnership will implement the SDGs (United Nations, 2015).

The First STI Forum was held in June 2016 and the last was the Seventh STI Forum held on 5 May 2021. In this Ministerial Session the theme was "Science, technology and innovation for building back better from the coronavirus disease (COVID-19) while advancing the full implementation of the 2030 Agenda for Sustainable Development" and the representatives of the countries tried to solve the following questions (STI Forum 2021):

- Best practices for equality

- STI4SDG roadmaps
- STI for resilience building
- STI towards low-carbon economies
- Lessons learned towards SDGs 4, 5, 14 and 15

The EU was the first regional integration organisation that aligned its policies and frameworks with the 2030 Agenda principles through the New EU Consensus for Development and the UE was the first to establish a Multi-Stakeholder Platform on SDGs. At the same time, the EU has also recognised the linkages between the 2030 Agenda and the Sendai Framework for Disaster Risk reduction, the Paris Agreement on Climate Change or the New Urban Agenda among other international frameworks (EU-CELAC Working Group, 2018).

At the inter-regionalism level, at the SOM EU-CELAC JIRI in 2020 the parties agreed to strengthen synergies with science, technology and innovation and multilateral mechanisms such as the STS, the Ibero-American Knowledge Space and the EU-LAC Foundation. On the basis of these regional mechanisms, projects can be developed that contribute to achieving the SDGs.

In 2022, the EU-LAC Foundation and the United Nations Development Program (UNDP) signed a strategic framework cooperation agreement to reinforce the integration and collaboration between the European Union and Latin America and the Caribbean towards the implementation of sustainable development goals. The agreement builds on the results of the 2021 dialogues on sustainable tourism which sought to foster bi-regional exchange of ideas among policymakers, entrepreneurs, civil society and researchers about conditions for moving towards a more sustainable, resilient and inclusive tourism sector that can contribute to the implementation of the Paris Agreement and the 2030 Agenda (UNDP, 2022).

As we can observe in the previous sections, there are projects, organisations and international for a where the implementation of policies to achieve the SDGs through multilevel governance are discussed, that is to say at national, regional and global levels. In most of these fora diplomats play an important role, firstly as facilitators in connection with the stakeholders and secondly as liaisons to monitor the initiatives or projects that may be undertaken.

The international organisations as the United Nations defined the SDGs. However, these goals can only be achieved if governments, academia, business and civil society work together. In the UN framework, there is the Major Group of other Stakeholders (MGoS). The EU and CELAC have a key role in the multilateral agenda and in the fulfillment of the 2030 Agenda. The concept of "development in transition" can serve as a boost for new forms of international cooperation for development (Sanahuja & Ruiz, 2020) and identify new cooperation areas in EU-CELAC inter-regionalism.

The 2030 Agenda, as well as the drive for a more sustainable energy transition, could be affected after the pandemics and the war between Russia and Ukraine. Ukraine is not the only country fighting to defend its independence. European countries have seen their energy independence threatened. The degree of dependence of European countries on products such as Russian gas or oil has shown that the solution is not provided by renewable energy, given that it would not currently solve the energy problem that Europe is going through. In fact, there are countries that are reconsidering reopening nuclear power plants and others building new plants. The degree of dependence of European industry on fossil fuels is so great that it is impossible to think that an economy based on renewable energies can be achieved in the coming decades.

In January 2022, the European Commission presented a report in which it granted "green" recognition to nuclear power plants that are already underway and those that are built until at least 2045. Gas-fired electricity generation plants will also enjoy the same recognition at least until 2030 (Euronews, 2022; El País, 2022). Undoubtedly, this change in approach will affect the objectives set out in the 2030 Agenda. Consequently, in the cooperation that the European Union develops with third countries and other regions in the world.

In the case of EU-CELAC relations, we will have to discuss again what kind of future we want to have. Many of the so-called "green" policies promoted by the European Union are now being questioned. Europe needs energy so as not to stop its industrial production and so that its citizens do not go cold during the winter, due to the increase in fuel prices, or worse still, the shortage of gas or oil.

Since the start of the war in February 2022, European countries have sought to replace their dependence on Russian supplies but have not yet succeeded. Algeria, Saudi Arabia, Norway and Venezuela are some of the alternatives to solve the gas and oil supply problem.

According to a report published by ECLAC, in April 2022 "Evolution of prices of natural resources for export from Latin America and the Caribbean", between 2020 and 2021, the prices of aluminium, copper, tin and iron grew by 42.8%, 48.5%, 86.1% and 46.1%, respectively. Among the arguments presented, it is pointed out that the increase in prices was due to the recovery of world economic activity; particularly to the increase in demand from China, a country that is the largest consumer of metals worldwide; and supply chain interruptions (CEPAL, 2022). Investment in infrastructure and construction, as well as growth in demand for automobiles, household appliances, and machinery, have been factors that have driven demand for metals (World Bank, 2021). Although it is true, it is considered that the short-term prospects for mining are positive due to the resurgence of world demand. However, there is uncertainty regarding the evolution of the pandemic and the war between Russia and Ukraine, facts that could affect the prices of some minerals such as titanium, palladium, neon gas, aluminium and nickel, products that have had a strong rebound in this period, due to the fact that Russia and Ukraine are two of the main world producers of the aforementioned products. Why are these minerals important? Because they are used to manufacture semiconductors, electric batteries and automobiles, three large economic sectors that are directly related to industry. Prior to the pandemic and the war in Ukraine, the energy transition was expected to increase demand for most metals, aluminium and copper (electric vehicles and infrastructure), nickel, cobalt and lithium (batteries) and rare earth metals (magnets for green technologies).

On the other hand, according to World Bank estimates (Chian & Baffes, 2021), electric vehicles require four times more copper than conventional cars. According to the IEA's long-range scenario (2021), lithium and cobalt consumption would have to increase more than sixfold to meet the needs of batteries and other clean energy uses, while copper use would double and that of nickel would quadruple. Finally, the report also indicates that according to ECLAC estimates (Jones et al., 2021), annual growth in demand for refined copper and lithium is projected to be 1.5% and 16.4% from 2019 to 2030, respectively. However, the IMF warns that the energy transition could be delayed by the increase in mineral prices (Boer et al., 2021).

Once again Latin America and the Caribbean are at a crossroads. Bet on unsustainable industrial development or bet on sustainable development. Empirical evidence shows that European countries are reconsidering their policies on renewable energy and fossil fuels, which will make them also reconsider their objectives with the 2030 Agenda. With the measures adopted by the European Commission, it is possible to continue thinking about reducing the levels of carbon dioxide in the atmosphere. Implement energy transition policies, from energy produced by fossil fuels to clean energy. Everything seems to indicate that for now they will continue betting on fossil fuels, which will consolidate the market for these fuels.

How can European countries require Latin American and Caribbean countries to develop sustainable development policies, to reduce their consumption of fossil fuels, if they themselves will not do so?

Undoubtedly, in this case, the European Commission has failed to respond correctly to the challenge posed by Russia. The evidence also leaves a clear reality for European society, discourse is one thing and reality is another.

It is interesting to see that this conflict could have served as an argument for betting completely on an energy change, therefore fulfilling the long-awaited "energy transition" by investing in renewable energies: solar, wind, Geothermal energy or tidal energy. Therefore, how can other countries be required to promote and implement sustainable energy policies if the main promoters, the European countries, fail to do so?

Undoubtedly, the transformation of the European energy system will be one of the main issues on the European Union's agenda in the coming years. And in this sense, natural resources and energy resources may be two topics within the future bi-regional agenda between the EU and CELAC.

Latin American and Caribbean countries have the necessary natural resources to be able to carry out real energy transition planning on both sides of the Atlantic, at the same time, they can help the European countries to solve their energy deficiencies.

There is also a possible scenario, in which China and Russia continue to consolidate their positions and economic and political influence in Latin American and Caribbean countries, and consequently, Europe is totally left behind.

There is always the option that Latin American and Caribbean countries assume their status as countries belonging to the Western world and therefore decide to strengthen their hemispheric relations with Canada and the USA and make a bi-continental alliance between Europe and America, This possible scenario reinforces the need to think in terms of a European Union-Organization of American States inter-regionalism. It must be taken into account that security and defence are crucial aspects in the development of a foreign policy focused on being a global player.

The Paris Agreement would not have been possible without the votes of Europeans, Latin Americans and Caribbeans. The current international situation makes it necessary to review the commitments assumed in the 2030 Agenda and the SDGs established as goals. We must think of a 2050 Agenda based on a reality, the transition to a green and sustainable economy will not be possible due to the dependence that countries have on fuels such as gas or oil. It is necessary to design a realistic plan that includes a gradual substitution and that includes large investments in renewable energies at a global level. The effort of a region is not enough, it is essential that all the regions of the world agree. Inter-regionalism demonstrates with this arduous and colossal task its importance and its role in the great issues of global governance.

## References

Boer, L., Pescatori, A., Stuermer, M., & Valckx, N. (2021). *Soaring metal prices may delay energy transition*. International Monetary Fund Blog – Insights and Analysis on Economics and Finance.

Brussels Declaration. (2015). *Shaping our common future: Working together for prosperous, cohesive and sustainable societies for our citizens*. European Council, Council of the European Union. Retrieved January 6, 2022, from https://www.consilium.europa.eu/media/23753/eu-celac-brussels-declaration.pdf

CEPAL. (2022). *Evolución de los precios de los recursos naturales de exportación de América Latina y el Caribe*. Retrieved January 6, 2022, from https://www.cepal.org/es/enfoques/evolucion-precios-recursos-naturales-exportacion-america-latina-caribe

Council of the European Union. (2017). *A sustainable European future: The EU response to the 2030 agenda for sustainable development*. Council conclusions. General Secretariat of the Council.

Cyted. (2018). *Estatuto y Reglamento Orgánico Programa Cyted*. Retrieved January 6, 2022, from https://www.cyted.org/sites/default/files/estatuto_y_reglamento_organico_cyted.pdf

Chian, K. W., & Baffes, J. (2021). *Sustained rebound in metal prices amid the global economic recovery*. Data Blogs World Bank.

ECLAC. (2022). *Fifth report on regional progress and challenges in relation to the 2030 agenda for sustainable development in Latin America and the Caribbean. A decade of action for a change of era*. Economic Commission for Latin America and the Caribbean.

ECLAC. (2021). *Social panorama of Latin America, 2020*. Economic Commission for Latin America and the Caribbean.

ECLAC. (2020a). *Special Report COVID-19*. N1 ECLAC, COVID-19 Response. Retrieved January 6, 2022, from https://repositorio.cepal.org/bitstream/handle/11362/45351/6/S2000263_en.pdf

ECLAC. (2020b). *The 2030 agenda for sustainable development in the new global and regional context: Scenarios and projections in the current crisis*. Economic Commission for Latin America and the Caribbean. Retrieved from https://repositorio.cepal.org/bitstream/handle/11362/45338/4/S2000207_en.pdf

ECLAC. (2016). *700 (XXXVI) Mexico Resolution on the establishment of the forum of the countries of Latin America and the Caribbean on Sustainable Development*. Santiago: Economic Commission for Latin America and the Caribbean. Retrieved January 6, 2022, from https://www.cepal.org/sites/default/files/document/files/700xxxvi-forum-ing.pdf

Elorza, A., Melchor, L., & Lacunza, I. (2021). *Nurturing the EU science diplomacy community: The launch of a EU science diplomacy Alliance for addressing global challenges*. S4D4C report. Retrieved January 6, 2022, from https://www.s4d4c.eu/wp-content/uploads/2021/03/Nurturing-the-EU-Science-Diplomacy-Community_EU-SD-Alliance.pdf

El País. (2022). *La Comisión Europea reconoce la energía nuclear como verde al menos hasta 2045*. Economía. Energía. Retrieved January 1, 2022, from https://elpais.com/economia/2022-01-01/la-comision-europea-reconoce-la-energia-nuclear-como-verde-al-menos-hasta-2045.html

European Commission. (2021a). *Horizon Europe, budget: Horizon Europe - the most ambitious EU research & innovation programme ever*. Directorate-General for Research and Innovation. Retrieved January 6, 2022, from https://op.europa.eu/en/publication-detail/-/publication/1f107d76-acbe-11eb-9767-01aa75ed71a1/language-en

European Commission. (2021b). *Horizon Europe – Investing to shape our future*. Retrieved January 6, 2022, from https://ec.europa.eu/info/files/horizon-europe-investing-shape-our-future_en

European Commission. (2021c). *Commission decision on the approval on behalf of the European Union of the EU-CELAC 2021–2023*. Strategic Roadmap for the implementation of the Brussels Declaration and EU-CELAC Action Plan on Science, Technology and Innovation. Retrieved January 6, 2022, from https://ec.europa.eu/info/sites/default/files/research_and_innovation/strategy_on_research_and_innovation/documents/eu-celac_strategic-roadmap-2021-2023.pdf

European Commission. (2018). *Science, research and innovation performance of the EU 2018. Strengthening the foundations for Europe's future*. Retrieved January 6, 2022, from https://op.europa.eu/en/publication-detail/-/publication/16907d0f-1d05-11e8-ac73-01aa75ed71a1/language-en

European Commission. (2012). *Communication from the Commission to the European Parliament, the Council, the European Economic and Social Committee and the Committee of the Regions. Enhancing and focusing EU international cooperation in research and innovation: A strategic approach*. Brussels. 14.9.2012 COM (2012) 497 final. Retrieved January 6, 2022, from https://www.eumonitor.eu/9353000/1/j9vvik7m1c3gyxp/vj2zs9r6zbvz

European Commission. (2015). *State of the innovation Union. Directorate-General for Research and Innovation* (p. 6). Luxembourg: European Union 2015.

European External Action Service. (2016). *Shared vision, common action: A stronger Europe*. A global strategy for the European Union's foreign and security policy. European External Actions Service, Brussels.

European Parliament. (2019). *The EU strategy for international cooperation in research innovation*. European Parliamentary Research Service (p. 9). Retrieved January 6, 2022, from https://www.europarl.europa.eu/RegData/etudes/BRIE/2019/631771/EPRS_BRI(2019)631771_EN.pdf

European Parliament and the Council of European Union. (2019). *Joint Communication to the European Parliament and the Council European Union, Latin America and the Caribbean: joining forces for a common future*. Retrieved January 6, 2022, from https://ec.europa.eu/international-partnerships/system/files/eu-lac-communication_en.pdf

European Union. (2021). *European External Action Service*. Delegation of the EU in Brazil. Science. Technology & Innovation Section. Retrieved from https://cdn5.euraxess.org/sites/default/files/news/he_wp_21-22_br.lac_.inco_final.pdf

EU-CELAC Working Group. (2018). *The future of the EU-LAC relations: Priority issues*. https://eulatnetwork.org/wp-content/uploads/2018/10/EU-CELAC-WG_PriorityIssues_EN_FIN.pdf

Euronews. (2022). Brussels unveils plan for "green" gas and nuclear label. *World News*. Retrieved January 6, 2022, from https://www.euronews.com/2022/01/01/brussels-unveils-plan-for-green-gas-and-nuclear-label

Gual Soler, M. (2020). *Diplomacia Científica en América Latina y el Caribe. Estrategias, mecanismos y perspectivas para fortalecer la diplomacia de la ciencia, tecnología e innovación.*

UNESCO. Retrieved from http://forocilac.org/wp-content/uploads/2020/11/PolicyPapers-DiplomaciaCientifica-ES.pdf

IEA. (2021). *Net zero by 2050 a roadmap for the global energy sector*. IEA.

InsSiDe. (2019). Building and interdisciplinary European science diplomacy. *InSciDe Brief*. Retrieved January 6, 2022, from https://www.insscide.eu/IMG/pdf/d9.7.pdf

Jones, B., Acuña, F., & Rodríguez, V. (2021). *Cambios en la demanda de minerales: análisis de los mercados del cobre y el litio y sus implicaciones para los países de la región andina"*. Documentos de Proyectos (LC/TC.2021/89). Repositorio CEPAL. Retrieved January 6, 2022, from https://repositorio.cepal.org/handle/11362/47136?show=full

Meyer, N. (Ed.). (2021). *The Impact Story of S4D4C*. S4D4C. Retrieved January 6, 2022, from https://www.s4d4c.eu/wp-content/uploads/2021/04/The-Impact-Story-of-S4D4C.pdf

Montevideo Declaration. (1992). *Inter-American Institute for Global Change Research*. Retrieved January 6, 2022, from https://www.iai.int/pdf/es/IAI_declaration_of_montevideo_sp.pdf

Quevedo, F. (2013). *The importance of international research institutions for science diplomacy*. Science 6 Diplomacy 2, N 3. Retrieved January 6, 2022, from https://www.sciencediplomacy.org/perspective/2013/importance-international-research-institutions-for-science-diplomacy

RICYT. (2022). *Políticas en Ciencia, Tecnología e Innovación en Iberoamérica*. Organismos Nacionales en Iberoamérica. RICYT (Red de Indicadores en Ciencia y Tecnología Iberoamericana e Interamericana). Retrieved January 6, 2022, from http://www.politicascti.net/index.php?option=com_weblinks&view=category&id=3%3Aorganismos-nacionales-en-iberoamerica&Itemid=6&lang=es

Santiago Declaration. (2013). *Council of the European Union*. Santiago de Chile, 27 January 2013. Retrieved January 6, 2022, from https://ec.europa.eu/commission/presscorner/detail/en/PRES_13_31

SDG Center for Latin America and the Caribbean. (2020). *Índice ODS 2019 para América Latina y el Caribe* (p. 21). Centro de los Objetivos de Desarrollo Sostenible para América Latina.

Sanahuja, J. A., & Ruiz, E. (2020). La Unión Europea y la Agenda 2030 en América Latina. Políticas de cooperación en una región de "desarrollo en transición". In J. A. Sanahuja (Ed.), *La Agenda 2030 en Iberoamérica. Políticas de cooperación y "desarrollo en transición"* (p. 353). Fundación Carolina.

Sebastián, J. (1992). The CYTED program: An innovative initiative for international cooperation in Ibero-America. *Impacto: Ciencia y Sociedad, 167*, 215–224.

S4D4C. (2019). *The Madrid Declaration on Science Diplomacy*. Madrid. S4D4C. Retrieved January 6, 2022, from https://www.s4d4c.eu/s4d4c-1st-global-meeting/the-madrid-declaration-on-science-diplomacy/

United Nations. (2021a). STI Forum. *Science, Technology & Innovation for the Sustainable Development Goals*. Seventh annual Multi-Stakeholder Forum on Science, Technology and Innovation for the Sustainable Development Goals. Note for Ministerial Session. Retrieved January 6, 2022, from https://sdgs.un.org/sites/default/files/2022-02/Note%20for%20Ministerial%20Session%20of%20the%202022%20STI%20Forum_final.pdf

United Nations. (2021b). *Training workshop for Latin America on policies and policy instruments*. Science, technology and innovation for the SDGs. United Nations, Department of Economic and Social Affairs. Sustainable Development. Retrieved January 6, 2022, from https://sdgs.un.org/events/training-workshop-latin-america-policies-and-policy-instruments-science-technology-and

United Nations. (2015). *Resolution adopted by the General Assembly on 25 September 2015*. Retrieved January 6, 2022, from https://www.un.org/ga/search/view_doc.asp?symbol=A/RES/70/1&Lang=E

United Nations. (1984). *United Nations Ibero-American program formulation document*. Ibero-American Program for Science, Technology and Development.

United Nations Development Programme. (2022). *UNDP and the EU-LAC Foundation renew their partnership to advance a common agenda for Sustainable Development*. Retrieved January 6,

2022, from https://www1.undp.org/content/brussels/en/home/presscenter/pressreleases/2021/undp-and-the-eu-lac-foundation-renew-their-partnership-to-advanc.html

World Intellectual Property Organization. (2021). *Global innovation index 2021. Tracking innovation through the COVID-19 crisis*. World Intellectual Property Organization.

World Bank. (2021). *World Bank Commodity Price Data (The Pink Sheet)*. Retrieved September 3, 2022, from https://www.worldbank.org/en/research/commodity-markets

# Chapter 8
# Science Diplomacy Between EU and CELAC

**Abstract** The chapter will focus on science diplomacy as a key cooperation area in science and technology between EU-CELAC. It will look at existing bi-regional projects and how these can be developed further in the future. The EU-LAC inter-regionalism has fostered a bi-regional cooperation based on science and Research and Development (R&D), it is time to evaluate how collaboration in science and technology between the EU and CELAC can advance towards an EU-CELAC science diplomacy, whose direct beneficiaries are the societies in both regions and whose indirect beneficiaries are global society. The development of an EU-CELAC science diplomacy policy can contribute global better governance and rebalance the international system.

**Keywords** Science diplomacy · Innovation · Technology · JIRI and EU-CELAC Knowledge Area

## 8.1 Science Diplomacy in Action: Science and Innovation in the EU-LAC Relations

Science, technology and innovation have become the real engines of social and economic progress in countries and the degree of investment in these areas in national budgets and the generation of policies that promote these areas represent the difference between technology-producing countries and technology consumers. Curiosity and interest in discovering new things and solving problems have allowed humanity to advance to new stages as a civilisation. Science has always offered solutions to the world's complex problems. In this world that is so independent, cooperation is not an option, it is the ideal way to achieve a peaceful and prosperous world.

Inequality between countries, like inequality between people, is mainly caused by the lack of opportunities. These opportunities begin with investment in education, knowledge and research. Science diplomacy is an instrument of international



M. Torres Jarrín, S. Riordan, *Science Diplomacy, Cyberdiplomacy and Techplomacy in EU-LAC Relations*, United Nations University Series on Regionalism 25, https://doi.org/10.1007/978-3-031-36868-4_8

cooperation that can help solve global problems. Hence the importance that countries attach in their development plans to foreign policies oriented towards scientific diplomacy.

Pandemics teach us the high degree of vulnerability of human beings. And although there was a certain degree of cooperation in the development of vaccines, such cooperation in response to the COVID-19 pandemic has been scarce. Rather we have seen increased geopolitical competition between countries, when surely the easiest thing would have been to combine human and financial resources to solve the pandemic, given its global characteristics.

Climate change forces countries to cooperate. Now two more changes are necessary, the energy change and the geopolitical change, in short, it is a change in the world order. Climate change and energy change require efforts in the private and public sectors to limit global temperature rise and stave off natural disasters.

The objective of the European Union to be climate-neutral by 2050 is questionable at this moment. Maybe the EU needs to reconsider the objectives of its European Green Deal which seeks to ensure: no net emissions of greenhouse gases by 2050, economic growth decoupled from resource use, no person and no place left behind. As part of the European Green Deal, the European Commission proposed on 4 March 2020 the first European Climate Law to enshrine the 2050 climate-neutral target into law. The law aims to ensure that all European Union policies contribute to this goal and that all sectors of the economy, governments and society play their part (European Climate Law, 2021):

- Setting the long-term direction of travel for meeting the 2050 climate neutrality objective through all policies, in a socially fair and cost-effective manner.
- Setting a more ambitious EU 2030 target, to set Europe on a responsible path to becoming climate-neutral by 2050.
- Creating a system for monitoring progress and taking further action if needed.
- Providing predictability for investors and other economic actors.
- Ensuring that the transition to climate neutrality is irreversible.

The Paris Agreement establishes requirements regarding the transparency of the commitments adopted, and there are a series of requirements around the enhanced transparency framework that include: use of transparency, accuracy, integrity, consistency and comparability principles.

In a study by the Economic Commission for Latin America and the Caribbean, in 2022, titled "Aumento de la ambición en la adaptación al cambio climático en América Latina" the document indicates that according to the Paris Agreement, each country must provide the following information, as appropriate, related to monitoring and evaluation (Bofill, 2022):

- Achievements, impacts, resilience, review, effectiveness and results.
- Approaches and systems used, and their results.
- Assessment of, and indicators to (measure):
  - How adaptation increased resilience and reduced impacts?
  - If adaptation is not sufficient to avoid impacts?

  - How effective are the adaptation measures implemented?
- Implementation, in particular on:
  - Transparency of planning and execution.
  - How to support programmes meet specific vulnerabilities and needs of adaptation.
  - How adaptation actions influence other development goals.

As well as good practices, experience and lessons learned from policies, regulatory changes, actions and coordination mechanisms. Coordination and joint work between the EU and CELAC in each of these areas is important and necessary to achieve compliance with the commitments assumed in the Paris Agreement.

In a changing and globalised world, where challenges affect every country, science and innovation cooperation can be the driving forces in the development of countries in economic, social and environmental terms. Climate change, food security, biodiversity, energy, poverty, natural disasters even pandemics, as COVID-19 has shown, are areas where science diplomacy can contribute to resolving and preventing future pandemics and therefore, to building a safe and better world.

Today, all countries are interconnected, and the problems of a country are regional or global issues. In this sense, diplomacy plays a crucial role as an instrument and mechanism for resolving international problems. For these reasons it is necessary that science diplomacy must be included at various political levels:

- Political parties
- Governments
- Parliaments
- Foreign services
- Academia

Science diplomacy can promote collaboration and cooperation between countries that foster harmony in international relations. On the one hand, the scientific sector requires international cooperation, and in this case, diplomacy serves as a platform to promote cooperation in science and innovation. On the other hand, in complex periods scientific cooperation can serve as a communication channel when diplomatic relations between countries are stagnant or broken. For the European Union: "science and technology play an increasingly important role in the geopolitical arena. Issues span from the race for technological supremacy between global powers to the increasing politicization, militarization, and commercialization of the global commons such as the polar zones, the high seas and outer space" (European Union, 2022a).

There are several platforms for science or technology cooperation between the European Union and other regions and, usually the European Commission organises and supports meetings between different stakeholders with a focus on science and technology.

In the framework of the European Union-Latin America and the Caribbean cooperation, we can find a strong tradition of exchanges and links between individual researchers from both regions. The LAC countries accelerated the process of institutionalisation and professionalisation of research centres thanks to European technical assistance, for example as LAC universities promoted hiring foreign researchers and granting scholarships.

The ways in which Science, research, innovation and technology constitute an independent chapter of the EU-CELAC Action Plan and the Senior Officials Meetings on Science and Technology (SOM) are examples of the EU-LAC scientific cooperation.

Since 2010 with the establishment of the EU-CELAC Joint Initiative on Research and Innovation (JIRI), the SOM has focused on implementing the EU-CELAC Knowledge Area. Since 2014 at the bi-regional research Project level, we can find INCO-Net projects (FP/EULARINET and ALCUE NET) and under the ERANET-LAC the first bi-regional call for science and technology was created. At bilateral levels, there are also joint calls, mobility schemes and scholarship programmes supported by the European Union through the ALFA and Erasmus Mundus+ programmes (FOCUS, 2022).

As we can observe there is a bi-regional cooperation acquis in science and innovation areas between European and Latin American and Caribbean countries. But this cooperation would be more fruitful if the countries had science diplomacy within their foreign policy and there were more specialised diplomats in this area. Although in practice there is a science diplomacy within the EU-LAC bi-regional agenda and cooperation.

## 8.2 Whether Collaboration in Science and R&D Really Amounts to Science Diplomacy, and How It Can Be Lifted to a More Policy Focused Level?

On the 28th and 29th of June 1999 under the Co-Presidency of the President of Brazil, Mexico and the Chancellor of Germany, in his capacity as President of the European Council the EU-LAC Strategic Partnership was launched in the framework of the First EU-LAC Summit held in Rio de Janeiro.

The Heads of State and Government decided to adopt the Declaration of Rio de Janeiro which included a section with the title “cultural, educational, scientific, technological, social and human fields”. In particular, the countries agreed to promote scientific research and technical development as fundamental elements for the successful insertion of countries in a globalised world, which requires the advances of scientific knowledge, as well as favouring innovation and technology transfers in order to obtain greater economic and technical linkage between the two regions in the fields of the production of goods and services, foreign trade, infrastructure and telecommunications and transport, among other issues (Declaration of Rio de Janeiro, 1999).

On 18th May 2010, the VI EU-LAC Summit was celebrated in Madrid with a specific chapter on science, innovation and technology and the Joint Initiative for Research and Innovation (JIRI) was established to enhance bi-regional dialogue on science, research and innovation. The governments highlighted the key role played by science, technology and innovation in achieving sustainable development and social inclusion. At the same time, this summit decided to strengthen the science, technology and innovation dialogue at the ministerial and EU-LAC S&T Senior Officials levels in order to ensure updating and monitoring of priorities and joint instruments and they engaged with the creation of the "EU-LAC Knowledge Area" (Madrid Declaration, 2010).

Since 2011 within the context of the JIRI five areas were identified, each of these areas cuts across societal challenges in different ways. In addition, a cross-thematic group was also established on "Good practices, funding mechanisms and Careers Development". The five Thematic SOM Working Groups target the integration or association of funded CELAC and EU research teams with national and/or regional policy agendas:

1. Bioeconomy (Argentina and France)
2. Biodiversity and Climate Change (Colombia and France)
3. ICT for meeting societal challenges (Chile and Finland)
4. Renewable Energies (Mexico and Spain)
5. Health (Spain and Brazil)
6. Cross-cutting "Good practices and funding mechanism" and "Careers Development" (Mexico and Portugal/Chile and Finland)

Among tasks of the JIRI Working Groups, the following areas were identified:

- Review of and recommendations for the annual Reports and activity plan prepared by JIRI Working Groups (JIRI WG) Leaders and Co-leaders.
- Elaborate proposals to be implemented by the JIRI WG.
- Identify synergies with ongoing related projects and regional and global initiatives in order to implement JIRI WG activities.
- Reflect on Horizon 2020 programme focused on upcoming and future calls and bi-regional cooperation.

The First EU-LAC Senior Officials "Meeting on Science and Technology" was held on 28th and 29th March 2011. To date, it is the only bi-regional mechanism that remains current and active, if considering that EU-LAC inter-regionalism created multiple dialogues at levels different as governmental, private sector, academia, civil society, cultural etc. During 2010 and 2011, Latin American and Caribbean countries created CELAC as a regional political forum and it was designated as a counterpart to continue the negotiations with the European Union. The issues of science, technology and innovation were addressed within the framework of the EU-CELAC Summit system.

In 2015 in the framework of the II EU-CELAC Summit the concept of an EU-CELAC Common Research Area was incorporated and its development established, focusing on three strategic pillars:

1. Mobility of researchers
2. Access to research infrastructures
3. Jointly addressing common global challenges

In parallel between 2007 and 2013 period the 7th Framework Programme supported bi-regional dialogues and projects as: ALCUE NET, ERANET-LAC, EULAC Health etc. ALCUE NET was a project conceived, according to the Brussels EU-CELAC Summit in 2015, with its main objective to establish a bi-regional platform for policymakers, research institutions and the private sector to design and implement policies and cooperation on Science, Technology and Innovation. ALCUE NET served to introduce concepts such as science diplomacy and strengthen and expand the National Contact Points in Latin American and Caribbean countries. The project was led by a Government Secretary of Science, Technology and Productive Innovation of Argentina and the period for the project was from 2012 to 2017.

ERANET-LAC was a bi-regional network on Joint Innovation and Research Activities that supported the political process of implementing the Joint Initiative for Research and Innovation (JIRI), which was endorsed by the Madrid Action Plan 2010–2012 adopted by the EU-LAC Summit in Madrid 2010. The project was led by the German Aerospace Center (Germany) and the project period was 2013–2017.

EU-LAC Health is a project with the aim of developing a common roadmap that enhances and coordinates the bi-regional collaboration between EU Member States and LAC countries in the health research area. The roadmap includes a prioritised Scientific Research Agenda defining the scientific objectives that allow the main global societal challenges to be addressed. The project was led by Carlos III Health Institute (Spain) and the project period was 2011–2017.

In March 2017, the EU-CELAC SOM JIRI established a Working Group on Research Infrastructure. The countries set goals for 2030 upgrading infrastructure and retrofitting industries to make them sustainable, enhancing scientific research and upgrading the technological capabilities of industrial sectors in all countries (ALCUE NET, 2017). During the 2014–2020 period, the European Program Horizon 2020 covered projects in areas such as climate change, research infrastructure, health, food security and transport, among others.

In the VIII SOM JIRI held on 30 October 2020 online and hosted by Mexico under the Pro-Tempore Presidency of CELAC, the participants underlined the importance of science and innovation cooperation on COVID-19, to ensure global access to medicines, vaccines and medical equipment, particularly in joining efforts to strengthen multilateral and bi-regional initiatives such as in the framework of the WHO, PAHO, GAVI and COVAX programmes. At the meeting participated 160 delegates from 40 countries. In a special session entitled "The world post COVID-19: Joint response to a global challenge", the participants highlighted initiatives promoted through other multilateral institutions and mechanisms that promote international cooperation in Science, Technology and Innovation, such as the STS Forum, the Ibero-American Knowledge Area and the EU-LAC Foundation. The meeting concluded with the consensus of the 2021–2023 Strategic Roadmap (JIRI, 2020).

The EU-LAC inter-regionalism has contributed to the countries of Latin America and the Caribbean developing national systems. We can say that in political terms, science, technology and innovation have been areas included in development policies, but these policies have not necessarily been endowed with budgets in accordance with the objectives that governments were intended to be achieved in productive performance and economic growth. The low participation of the private sector or the lack of public–private alliances in research and development projects are proven facts. In other words, we can say that economic growth seems to be dissociated from the levels of investment in science, technology and innovation (Cimoli, 2000; Dutrénit & Sutz, 2014).

To resolve the shortcomings described above, it is necessary to develop an innovation model of knowledge-based economies integrating and differentiating forces such as governmental agencies (public sector), research centres (academia sector) and business sector. In other words, apply the innovation model known as Triple Helix: university–industry–government relations (Etzkowitz & Leydesdorff, 1995). There is also an innovation model that includes Civil Society, which is known as the Quadruple Helix (Leydesdorff, 2012). According to Etzkowitz (2008) the Triple Helix aims to increase the benefits from national, regional, and multinational systems. On the other hand, this form of cooperation provides a balance between knowledge, social benefit and profit (Asheim & Coenen, 2005; Leydesdorff, 2005).

Since 2000, the European Union has been using the Triple Helix model to face the challenges of the Lisbon Strategy, which promoted the "information and innovation society" and "society based on knowledge" (European Parliament, 2000). The European Commission tried to confront the new challenges of its economic role in the global economic arena. During the current decade the European Union has prepared a new agenda for sustainable development and its intent is to comply with its international compromises and highest environmental standards to involve and coordinate all the stakeholders in this process, for which reason the new agenda goals include a Quintuple Helix approach (Franc & Karadzija, 2019).

During the last seven decades, academia has theorised about the development models in countries, mainly in developed countries, which focused on their production and industrialisation of goods and services using technology and innovation. According to Meissner and Kotsemir (2016) and Ranga (2020) throughout these last decades we can identify different models of innovation which we can divide into five periods:

1. First period (1950s–1970s): Linear models (Bush, 1945; Rogers, 1962). Science push-market pull.
2. Second period (1980s): Interactive models. Simultaneous coupling model (Rothwell & Zegveld, 1985): Chain-linked mode (Kline & Rosenberg, 1986).
3. Third period (1980s–2000s): National Innovation Systems (Freeman, 1987; Lundvall, 1992; Nelson, 1993; Edquist & Lundvall, 1993). Regional Innovation Systems (Cooke et al., 1998).
4. Fourth period (1990–2000s): Triple Helix (Etzkowitz, 1993, 2002; Etzkowitz & Leydesdorff, 1995).

5. Fifth period: Quadruple/Quintuple/N-tuple Helix (Yawson, 2009; Carayannis & Campbell, 2012; Leydesdorff, 2012).

Given that EU-CELAC relations are a multiplayer inter-regionalism (Torres Jarrín, 2018) because they includes multiple stakeholders in the bi-regional dialogue, EU-LAC countries can design a joint bi-regional innovation model that allows them to claim a space within the global economy.

## 8.3 The Role of Science Diplomacy in the EU, LAC, Ibero-American and American Regions

The distinctions between science, scientific, technological or innovation diplomacy are not clear, but there is broad agreement on the link between science policy and foreign policy (Leijten, 2017; Langenhove, 2017). The European External Action Services (EEAS) works closely with several European scientific institutions such as the European Commission's Joint Research Centre (JCR) and agencies like the EU Institute for Security Studies (EUISS) and the EU Satellite Centre (SatCen). Through them, it gathers information and gets experts who can advise EEAS and other EU institutions.

Various EU delegations count on the support of research and innovation counsellors seconded by the European Commission. The establishment of a Science and Technology Advisor position in the Policy Planning and Strategic Foresight Division has strengthened further the role of science at the EEAS. Since 2021, the EEAS has promoted an informal Network of Science Advisors and Science Diplomacy Coordinators in EU Ministries of Foreign Affairs aimed at providing a platform for facilitating the exchange of experiences and best practices among diplomatic services on science and technology-related matters (European External Action Service, 2022a). Science diplomacy has the potential to promote cooperation and conflict prevention, rebuild trust and foster shared understanding across countries and regions (European External Action Services, 2022b).

In May 2020 the S4D4C project launched the European Science Diplomacy Online Course. So far more than 6500 participants have started the course (European Union Science Diplomacy, 2022). This good practice makes two points: first the connection between science and diplomacy in international affairs; second, the importance of science in the foreign policy of the EU Member States. From 22 to 24 March 2022, the InsSciDE's Open Conference was held in Lisbon, under the title "European Science Diplomacy, Diversity and the Global South". The conference had two aims (European Union Science Diplomacy, 2022):

- To continue the debate on European Science Diplomacy, its collaborative endeavours as well as the tensions among countries, and to discuss the kaleidoscopic formats of the concept of technoscientific diplomacy, both time and geography-wise, and the actors involved in its practices.

- To explore new avenues of collaboration with non-European researchers and practitioners, particularly from the so-called global South.

In 2019, a Webinar on "Diplomacy of Innovation and Science Diplomacy: A vision for EULAC" was organised in the EULAC Focus project framework. As we mentioned before there are many projects, programmes and networks on science, technology and innovation, but there is no bi-regional platform that includes science advisors. LAC countries do not have any particular or specific regional integration organisation to create a science diplomacy policy. It would be better for both regions (EU and LAC) to have better science diplomacy policies.

The COVID-19 pandemic and the irruption of new technologies in societies represent a major challenge to EU and LAC countries. The pandemic showed the importance of international collaboration and scientific networks. In this sense, we can confirm that there are mechanisms and instruments that promote bi-regional cooperation on science, technology and innovation. According to ECLAC the COVID-19 consequences for Latin American and the Caribbean countries will increase by at least 4.4 percentage points the poverty levels in the whole region. This represents an average of around 28.7 million additional people compared with the poverty levels prior to the pandemic (ECLAC, 2020).

On 28 February 2022, the Ibero-American Secretariat organised the I Ibero-American Forum on Science Diplomacy. The aim was to establish a Network of Science Diplomacy that designed national strategies able to take advantage of the Community of Ibero-American countries. Andrés Allamand, Ibero-American Secretary, indicated the need to have a common action plan on Science Diplomacy (SEGIB, 2022). The interest of Ibero-American countries in science diplomacy was expressed during the III Meeting of Ministers and High Authorities of Science, Technology and Innovation held in Guatemala on 29 and 30 October 2018. The meeting was led by the National Council for Science and Technology of Guatemala with the support of the Ibero-American General Secretariat. Spain, a country that had been working for the last decade to incorporate science diplomacy among the priorities of Spanish foreign policy, offered its experience and best practices in science diplomacy to coordinate these actions.

The Andorra Declaration pointed out that the pandemic has demonstrated the need to strengthen national systems of science, technology and innovation, ensuring investment in Innovation and Development (R&D), promoting cooperation between countries and the transfer of knowledge and innovative technologies to developing countries (Andorra Declaration, 2021). At the same time, the Ibero-American countries approved the "Ibero-American Innovation Strategy" which will allow the promotion, through science, technology and innovation, of socioeconomic recovery and the range of goals set forth in the 2030 Agenda. The EU is committed to contributing to an effective cooperation with LAC region to achieve the SDGs Agenda 2030 (Ruiz & Sanahuja, 2019).

From an American perspective, there is a continental space dedicated to promote the science and technology cooperation—we refer to the cooperation developed within the framework of the system of the Organization of American States (OAS): Meeting of Ministers and High Authorities of Science and Technology,

Inter-American Network of Competitiveness, Prospecta Americas, Hub on Technology and Commercialization for the Americas (OAS, 2022).

On 7 and 8 December 2021, the OAS and the Government of Jamaica organised the VI Meeting of Ministers and High Authorities of Science and Technology under the theme "Harnessing the power of transformative science and technologies to drive our communities forward". The ministerial meeting addressed the following themes (Government of Belize, 2021):

- Science and technology to build resiliency in the framework of sustainable development and COVID-19 recovery.
- Youth: Improving skills and readiness for Industry 4.0.
- Science, technology and entrepreneurship to reduce the digital divide and promote the inclusion of women and girls, rural and indigenous communities, and other populations in vulnerable situations.
- Effective public–private–academia collaboration to enhance competitiveness and quality of life.
- Levelling the playfield for active participation in the global economy by micro, small and medium enterprises through STI.
- Technology foresight as input for public policy decisions (Prospecta Americas).
- Science for decision-making.

During the meeting, the authorities exchanged good practices and identified offers for cooperation to help implement hemispheric actions to accelerate the recovery from the pandemic of COVID-19.

Prospecta Americas is a programme promoted by the OAS whose objective is to share good practices and experiences among regional experts in future technologies, support the development of capacities, generate joint collaboration projects and improve the social appropriation of knowledge of 10 transformative technologies for the region (Prospecta Americas 2021):

1. Big Data
2. Internet of Things (5G)
3. Virtual reality
4. Artificial intelligence
5. Quantum computing
6. Robotics
7. Gene editing
8. Biomedical engineering
9. 3D and 4D manufacturing
10. New nanostructured materials

The Prospecta programme was launched at the first seminar on Technology Prospective for the Americas in Lima on 24 and 25 October 2019. The last such seminar was celebrated in Hidalgo, Mexico on 18 and 19 May 2022.

On 30 October 2020, EU and CELAC endorsed the outcome of the VIII Senior Officials Meeting on Science Technology (SOM) reviewing the developments in the EU-CELAC Common Research Area in each of the three pillars: mobility of researchers, cooperation of research infrastructures and global challenges. The countries approved the inclusion of a section called "Digital" that indicated: "the participants agreed that connectivity is an essential enabler of knowledge exchange, productivity and growth in this digital age and welcomed the efforts by both sides to establish a high-speed link between both regions". At the same time, the participants considered "cybersecurity as a vital element in building trust in the digital environments, helping secure the novel IT-based economy and digital lives of the citizens (European Commission, 2021).

The EU-CELAC initiative as Education Common Area, Research Common Area and the Ibero-American Space of Knowledge can contribute to promoting a Science Diplomacy in Action and comply with its goal of fostering harmony in international relations.

## 8.4 How Collaboration in Science and Technology Between the EU and CELAC Can Advance Science Diplomacy, Strength the Multilateralism, Benefit Each Other's Societies and Contribute to a Global Governance Balance and Better?

International trade agreements need to adapt to the Fourth Industrial Revolution because negotiations are lagging behind the breakneck progress of technological change (Inter-American Development Bank, 2017). In this sense, integration policies may play a part in creating innovation clusters (Intal, 2016) where EU-CELAC cooperation with a focus on science diplomacy can contribute to elaborating better regional policies.

The European Union-Community of Latin American and Caribbean States (EU-CELAC) cooperation includes a Science and Technology component. The EU is pushing for science diplomacy towards Latin American and Caribbean countries at various levels and in numerous projects and initiatives (Selleslaghs, 2017). There are several agreements with specific countries on science and technology: Argentina, Brazil, Chile and, Mexico (Sanchez, 2018).

The digitalisation process has transformed the economy, business, governments and societies. Just as the United Nations defined the 17 SDGs, it has likewise identified 11 norms of responsible state behaviour in cyberspace:

1. Interstate cooperation on security
2. Consider all relevant information
3. Prevent misuse of ICTs in your territory
4. Cooperate to stop crime and terrorism
5. Respect human rights and privacy
6. Do not damage critical infrastructure
7. Protect critical infrastructure

8. Respond to request for assistance
9. Ensure supply chain security
10. Report ICT vulnerabilities
11. Do no harm to emergency response teams (CERTS)

No doubt that cyber security is one of the major challenges facing all states in the modern era and cyberspace is now at the heart of international geopolitical tensions (Federal Department of Foreign Affairs of Switzerland Confederation, 2021). The international system requires cyberspace governance, and these efforts require the collaboration of stakeholders: countries, regional integration organisations and tech industry companies (Torres Jarrín, 2021). In this sense, the collaboration between EU-CELAC can contribute to establishing international norms for cyberspace.

From an international relations perspective, multilateralism is the joint action of several countries on a particular issue, instead of doing it unilaterally. The purpose of multilateralism was to abandon unilateral practices by some states against others. The unilateral actions created armed conflicts and wars. Hence, after two world wars, countries opted to promote peace through multilateralism.

The Member States of the European Union and the Member countries of the Community of Latin American and Caribbean States are defenders of multilateralism as an instrument and mechanism for solving international problems and confronting global challenges like poverty, inequality, security, climate change, among others and lately, the pandemic caused by COVID-19. However, the current armed conflicts in different regions around the world, the war between Russia and Ukraine, the political tensions between the United States and China, the return of powers such as Japan, Turkey and Iran or the rise of emerging powers such as India, Saudi Arabia, Indonesia, Brazil, Russia, Mexico, Nigeria and South Africa are examples of the complex balance of power within the international system and future world order. If we add to this that each continent has several regional integration organisations, then the number of international actors' multiples. Without considering, the Big Tech Companies that are the new geopolitical actors and that are the biggest promoters and producers of science and technology whose innovations are transforming societies, markets and international standards.

The pandemic caused by COVID-19 has exposed the weakness of the countries' health systems, European, Latin American and Caribbean, and has also shown that societies, governments and companies are not prepared for the era of digitalisation. The main international organisations have been talking for decades about the importance of strengthening health and social systems, investing in research, developing pandemic prevention systems, as well as developing societies based on knowledge, science and technology. All this requires greater investment in education and innovation. Today digital skills are the greatest asset that a person can have to find a job or have the greatest potential to create companies. However, governments have not allocated the necessary resources, nor have they created regulations for digital times.

During the pandemic, several countries tried to strengthen their influence through the supply of vaccines, which showed the return of unilateralism as a practice in international relations. The Latin American and Caribbean countries were not able to coordinate vaccine purchases, despite having several integration organisations at both the subregional and regional levels.

On the one hand, powers such as the USA, China, Russia and on the other hand, the European Union, sought to strengthen their presence in regions where they had some influence or in other cases recover the influence lost in recent years. The truth is that China and Russia managed to increase their influence in Latin American and Caribbean countries compared to the USA and European Union. Moreover, China is the main trading partner of several Latin American and Caribbean countries and assists these countries with technical cooperation and loans. We can therefore conclude that China's strong presence is to the detriment of European countries, which had earlier replaced the USA as the region's main trading partner and supplier of development aid. But those times have changed.

For their part, Latin America and Caribbean countries have also opted to strengthen their relations with Asian countries, rather than with European ones. The relationship with the USA continues to be a cold and distant relationship on both sides, that is from the USA and Latin American and Caribbean governments.

We are experiencing a phase change in international relations. Multilateralism is in crisis and global governance lacks leadership. The world order needs to be rethought. In this complex international scenario, inter-regionalism is called to play an important role in seeking a new balance of power within global governance. In this sense, the EU-CELAC inter-regionalism can be good practice and a model to establish a balance of power in the international arena.

The international press has not paid much attention to a series of initiatives that have strong impact on international relations, and consequently in EU-CELAC inter-regionalism: The "Alliance for Multilateralism", "The Summit for Democracy", "Eurasian Economic Union" and the "Belt and Road Initiative". Prior to a Sino-Russian meeting, Sergéi Lavrov, Russian Minister of Foreign Affairs, indicated that the world was "living through a very serious stage in the history of international relations. A new world order". Lavrov said: "We, together with you, and with our sympathizers will move towards a multipolar, just, democratic world order" (CBS, 2022). The Chinese Foreign Minister Wang Yi said that "Moscow and Beijing will continue efforts to advance global multipolarity and the democratization of international relations". The Chinese president, Xi Jinping mentioned "China and Russia cooperation has no limits" and "our striving for peace has no limits, our upholding of security has no limits, our opposition towards hegemony has no limits" (Devonshire-Ellis, 2022). It should be remembered that in June 2021 both countries renewed their alliance through the Treaty of Good-Neighborliness and Friendly Cooperation.

On 04 February 2022, the Sino-Russian Summit was held. After the meeting, the countries published a "Joint Statement of the Russian Federation and the People's Republic of China on the International Relations entering a New Era and the Global Sustainable Development (Effective)". The document defines Russia and China as "world powers": "The sides note that Russia and China as world powers with rich cultural and historical heritage have long-standing traditions of democracy, which rely on thousand-years of experience of development, broad popular support and consideration of the needs and interests of citizens. Russia and China guarantee their people the right to take part through various means and in various forms in the administration of the State and public life in accordance with the law. The people of both countries are certain of the way they have chosen and respect the democratic systems and traditions of other States" (Joint Statement of the Russian Federation and the People's Republic of China on the International Relations entering a New Era and the Global Sustainable Development [Effective], 2022).

In a European Parliament's document it is noted that "Russia and China criticize AUKUS and 'the formation of closed block structures and opposing camps in Asia Pacific'. Russia declares 'its support for the One-China principle, confirms that Taiwan is an inalienable part of China and opposes any forms of independence of Taiwan'. The two countries voice their opposition to NATO enlargement, and China gives its support to Russia's 'proposals to create long-term legally binding security guarantees in Europe" (European Parliament, 2022a).

The new alliance between Russia and China whose objective is to establish a new world order shows that we face an uncertain world. The Chinese "One Belt One Road" initiative—a global infrastructure development strategy, whose trade network included Asian, African, European, Latin American and Caribbean countries—is reconfiguring the old-world order. In Europe, it is feared that China will consolidate its supremacy in the world and divide Europe, but it is the same case in the American continent.

The world has known a multilateralism led by Western countries, mainly by the USA and some European countries, but everything seems to indicate that the future presents us with an Asian Multilateralism, led by China. For now, the world has begun to accept the coexistence of democratic countries with undemocratic and authoritarian countries in the main international forums and regional organisations. In the Latin America and Caribbean region, we find the cases of Cuba, Venezuela and Nicaragua. China and Russia expose that there are different forms of democracy, and that these countries are willing to defend their own conceptions of democracy.

In December 2021, the Information Office of the State Council of China published a White Paper entitled "China: Democracy that works". The document indicates that democracy is a common value of humanity and an ideal that has always been cherished by the Communist Party of China and the Chinese people. It also mentions that for the past hundred years, the Party has led the people in realising people's democracy in China. The document has a section titled "Promoting Democracy in International Relations" that states: "China proposes to build a global community of shared future, and presses for a new model of international relations based on mutual respect, fairness, and justice, and win-win cooperation. At the same time of intense global competition in the economic, scientific, technological, and other fields, China regards other countries as partners rather than rivals, it does not engage in cold war, confrontation, control or manipulation, but rather promotes mutually beneficial exchanges and cooperation" (The State Council Information Office of the People's Republic China, 2021). It is a model of socialist democracy that encompasses all aspects of the democratic process and all sectors of society. The document points out that this is a real democracy that works. There is no doubt that China plays and will play an important role in EU-CELAC inter-regionalism.

European, Latin American and Caribbean countries are in a dilemma. On the one hand, both the USA and China are their main partners. The transatlantic alliance is one of the main pillars in the foreign policy of most of the Member States of the European Union, however, they are aware that the centre of the world has turned towards Asia, and that in this new centre there is a nucleus called China, and that it is necessary to establish a privileged relationship as strong as with the USA. Latin Americans and Caribbeans have the same dilemma. The difference with the Europeans is that the former does not have a body that coordinates foreign policies and trade policies, which are necessary when negotiating with a great power such as China.

In 2014, the CELAC-China Forum was created, since then investment in infrastructure, trade and intergovernmental loans by the Chinese government to Latin American and Caribbean countries have been a constant. In 2021, CELAC and China signed an agreement that included strategic areas to be developed until 2024. The year 2024 is established as the objective to celebrate a new edition of the CELAC-China Forum. Among the defined areas of cooperation are bilateral and multilateral agricultural cooperation; the strengthening of scientific, technological and academic exchanges; as well as promoting economic development and the well-being of people.

On 21 February 2022, China and CELAC established an Action Plan, in which China committed to increasing cooperation in digital infrastructure, telecommunications equipment, 5G, Big Data, cloud computing, artificial intelligence, Internet of Things and smart cities, among others, in addition to exploring the construction of joint laboratories. Likewise, it indicates that according to studies by the Inter-American Development Bank (IDB), the region needs to invest 3.1% of its GDP in infrastructure annually to meet its Sustainable Development Goals for 2030. Of which 59% should be allocated to new infrastructure, and the rest for maintenance and replacement of existing assets. China plans through its Belt and Road Initiative to increase its investment in infrastructure (China-CELAC Action Plan, 2022).

Integration in Latin America and the Caribbean is more than ever necessary. If CELAC is the current regional mechanism that brings together all the countries, then they must strengthen it. It should begin by establishing its legal personality, competencies and creating supranational organisations that oversee designing and implementing regional and interregional policies.EU-CELAC inter-regionalism is vital for both regions. Both are weak compared to the other great powers. The international system seems destined to experience a phase in which there will be a division between countries that defend a "liberal democracy" following the Western model, and other countries that choose to follow a model of democracy led by China. China also has the risk that the West strengthens itself and prevents China from consolidating itself as the leading world power. A transatlantic alliance whose axis is made up of all the countries of the Western Hemisphere and Europe could be a significant counterweight to China and could tip the balance towards an empowerment of the Western world.

The history of European integration is linked to Latin American integration because both regions share a common history, the history of the West. Sometimes, some Latin American and Caribbean governments criticise the West, as if they were not part of it. However, the West is made up of both the continents of Europe and America. Consequently, the defence of the Western model and its future is also the responsibility of Latin American and the Caribbean. The decline, reform or empowerment of the West depends on Europeans and Americans, including within Americans the countries of North America, Central America, the Caribbean and South America.

Multilateralism and democracy seem to be in question. Both systems are Western.

The world order has been led by Western countries. The non-participation of Latin American and Caribbean countries in decision-making on issues on the international agenda has been their option, as the option is wanting to defend the Western model of multilateralism and democracy. Without Latin American and the Caribbean, the USA and the European Union can do little. If an agreement is still not reached, then it is likely that we will begin a new stage in the history of international relations based on multilateralism and democracy on the Eastern model, mainly led by China. Like the economy, history is usually made up of cycles. We should recall that Christopher Columbus was looking for a better route to the Far East when he discovered America. And that at that time the great power was China.

We should also recall that many of the countries today called emerging powers were centuries ago great powers: the Chinese Empire, the Persian Empire, the Russian Empire, the Ottoman Empire or the different empires developed in the Indus Valley. These are historical examples, which should be taken into account. The defence of cultures and the role of civilisations in the history of humanity are arguments on which governments often design their foreign policy. The manifest destiny of the role they play in the world is a constant variable in most of the histories of civilisations. The UK itself, by separating from the European Union, seeks to recover its legacy as a great world power. Therefore, as the Chinese and Russians rightly state, we are witnessing a moment of change of stage in the history of international relations. It will be up to the Latin American and Caribbean countries to decide if they want to be observers of global changes or global actors.

When the EU established a strategic partnership with Latin American and the Caribbean countries in the First bi-regional Summit, held in Rio de Janeiro (Brazil) in 1999, it did so on the basis of strong historical, cultural and economic ties. The EU defines the Latin American and Caribbean regions as a Global Actor, and their expectations were to lead global governance together. Together, EU and LAC countries account for more than a third of the UN membership and are a force for a strong and rules-based multilateral order. EU-LAC inter-regionalism is based on an international agenda of shared interests that includes global issues such as sustainable development, climate change and protection of biodiversity, human rights and fair and free trade (European Union, 2022b). The Declaration of the CELAC Summit held in Mexico on 18 September 2021, stated that CELAC will seek to "contribute to the strengthening and positioning of Latin America and the Caribbean, in the current regional and global political context, and continue speaking out as a region in the United Nations and other multilateral bodies" (Declaration of Mexico, 2021).

The European Union, Latin America and the Caribbean are also affected by the negative consequences of the Russo-Ukrainian war, for example rising prices and global shortages of food, fertilizers and energy, despite the LAC region being one of the largest producers of food in the world.

In 2022, Argentina held the Pro-Tempore Presidency of CELAC. The III EU-CELAC Foreign Affairs Ministers Meeting was held in Buenos Aires on 27 October 2022. Among the main topics to be discussed were two issues that Argentina has been raising in different international forums: improve the functioning of the international financial architecture, including the guidelines for the general allocation of Special Drawing Rights of the International Monetary Fund, and the need for the agency to carry out a review of its surcharge policy (Ministry of Foreign Affairs, International Trade and Worship of Argentina, 2022). Relaunching the EU-CELAC summits and talking about relations with Russia and China were also items discussed on the bi-regional agenda.

For Josep Borrell, High Representative of the European Union for Foreign Affairs and Security Policy and Vice-President of the European Commission, the LAC "region is an important partner for a very important number of reasons, and we agreed to step up what we can call the 'other transatlantic relationship'. Because when we talk about the transatlantic relationship, everybody thinks about the relationship with the United States. There is another transatlantic relationship, which is the relationship with Latin America and the Caribbean" (European Union External Action, 2022).

On 28 August 2022, the Spanish daily *El Pais* published a leaked European External Action Service document indicating the growing economic and political weight of China and Russia in LAC countries. The text mentioned that while the EU has focused more on neighbouring countries like Libya, Syria and Ukraine over the last decade, China increased its investment in LAC region around 26-fold during 2000–2020 period. Likewise, currently 21 LAC countries have signed onto the Chinese Belt and Road Initiative. The LAC region has the majority of the world's know lithium, petroleum and natural gas deposits. On the other hand, the document indicates that although Latin American and Caribbean countries were more aligned with the West countries in the United Nations votes condemning the Russia war, in the latest XIV Euro-Latin America Parliamentary Assembly, held in Buenos Aires, from 11 to 14 April 2022 (European Parliament, 2022b), the European parliamentary delegation was unable to pass a condemnation of Russia (De Miguel, 2022; McMurtry, 2022). The EEAS document also noted a trend of new "anti-establishment" leaders in Peru, Chile, Colombia, Honduras and Costa Rica. The triumph in the recent Brazilian elections of Lula Inácio Lula da Silva has consolidated this "remarkable transition" from right to left. Finally, the document also noted that the European Union is studying a package for 2023 that could mobilise up to 8 billion euros of European investment in the LAC region. In this way, the EU seeks to increase its power and leverage in LAC and, at the same time, recover the EU-LAC strategic partnership.

During the Spanish presidency of the Council of the European Union in the second half of 2023, the system of EU-CELAC Summits was restored and with it the interregional cooperation will continue, and agreements be approved (EU-Mexico, EU-Chile and EU-MERCOSUR), as well as common positions on China and Russia.

It remains to be seen if Latin American and Caribbean countries renounce their trade relations, investments and defence agreements with Russia and China, restore relations with the European Union, join the transatlantic axis (European Union-United States) or seek their own way, a third way where they are a counterbalance in the struggle for power and hegemony in the international system. CELAC-EU inter-regionalism faces a difficult test. The cooperation in the areas of science and technology can be the pillars on which the countries can redesign an agenda that was only focused on trade and investment. Both regions can strengthen or weaken the existing multilateral system. Otherwise, both regions may be relegated to international ostracism, to be mere irrelevant actors in the new world order that is being reconfigured.

## References


Andorra Declaration. (2021). *Declaración de Andorra. Innovación para el Desarrollo Sostenible – Objetivo 2030.* Iberoamérica frente al reto del Coronavirus. XXVII Cumbre Iberoamericana de Jefes de Estado y de Gobierno de Andorra., 21 de abril de 2021. Madrid. Secretaria General Iberoamericana. Retrieved August 21, 2022, from https://www.fundacioncarolina.es/wp-content/uploads/2021/11/Declaracio%CC%81n-XXVII-Cumbre-Andorra-ES.pdf

Asheim, B. T., & Coenen, L. (2005). Knowledge bases and regional innovation systems: Comparing Nordic clusters. *Research Policy, 34*(8), 1173–1190. Retrieved August 18, 2022, from https://www.sciencedirect.com/science/article/abs/pii/S0048733305001101?via%3Dihub

ALCUE NET. (2017). European Union and the community of Latin American and Caribbean states. 6th Senior Official Meeting on Science and Technology, Brussels, 14–15 March 2017. Retrieved May 12, 2022, from http://alcuenet.eu/policy.php

Bofill, P. (2022). *Aumento de la ambición en la adaptación al cambio climático en América Latina.* Documentos de Proyectos (LC/TS 2022/7) Santiago. Comisión Económica para América Latina y el Caribe.

Bush, V. (1945). *Science – The endless frontier. A report to the president.* United States Government Printing Office.

Carayannis, E. G., & Campbell, D. (2012). *Mode 3 knowledge production in quadruple helix innovation systems: 21st-century democracy, innovation, and entrepreneurship for development.* Springer.

CBS. (2022). Russia says it's building a new "democratic world order" with China. 30 March 2022. Retrieved March 30, 2022, from https://www.cbsnews.com/news/russia-china-lavrov-visit-beijing-vladimir-putin-xi-jinping-new-world-order/

China-CELAC Action Plan. (2022). *El Plan de Acción China-CELAC.* 21 de febrero de 2022. Sistema Económico Latinoamericano y del Caribe. Documentación. Retrieved August 21, 2022, from https://www.sela.org/es/centro-de-documentacion/base-de-datos-documental/bdd/79068/el-plan-de-accion-china-celac

Cooke, P., Heidenreich, M., & Braczyk, H.-J. (1998). *Regional innovation systems. The role of governance in a globalized world.* Routledge.

Cimoli, M. (Ed.). (2000). *Developing innovation system, Mexico in the global context.* Pinter Series.

Declaration of Rio de Janeiro. (1999). *European Parliament.* Retrieved May 12, 2022, from https://www.europarl.europa.eu/intcoop/eurolat/key_documents/summits_eu_alc/i_29_6_1999_rio_en.pdf

Declaration of Mexico. (2021). *Declaración de México, VI Cumbre de Jefes de Estado y de Gobierno de la Comunidad de Estados Latinoamericanos y Caribeños.* Ciudad de México, 18 de septiembre de 2021. Sistema Económico Latinoamericano y del Caribe. Documentación. Retrieved August 18, 2022, from http://www.sela.org/es/centro-de-documentacion/base-de-datos-documental/bdd/76182/declaracion-de-ciudad-de-mexico

De Miguel, B. (2022). *Bruselas prepara una ofensiva comercial y diplomática para frenar el avance de China y Rusia en Latinoamérica.* El País. Internacional. Unión Europea. Retrieved August 18, 2022, from https://elpais.com/internacional/2022-08-18/bruselas-prepara-una-ofensiva-comercial-y-diplomatica-para-frenar-el-avance-de-china-y-rusia-en-latinoamerica.html

Devonshire-Ellis, C. (2022). China, Russia stand firm in "new, fair, multipolar world order". *China Briefing.* Retrieved March 30, 2022, from https://www.china-briefing.com/news/china-russia-stand-firm-in-new-fair-multipolar-world-order/

Dutrénit, G., & Sutz, J. (Eds.). (2014). *National innovation systems, social inclusion and development: The Latin American experience.* Edward Elgar.

ECLAC. (2020). *El desafío social en los tiempos del COVID-19. Informe especial N3.* CEPAL.

Edquist, C., & Lundvall, B.-A. (1993). *Comparting Danish and Swedish systems of innovation.* Oxford University Press.

Etzkowitz, H., & Leydesdorff, L. (1995). The triple helix. University-industry-government relations: A laboratory for knowledge based economic development. *EASST Review, 14*(1), 14–19.

Etzkowitz, H. (2008). *The triple helix: University-industry-government innovation in action.* Routledge.

Etzkowitz, H. (2002). *MIT and the rise of entrepreneurial science.* Routledge.

Etzkowitz, H. (1993). Technology transfer: The second academic revolution. *Technology Access Report, 6,* 7–9.

European Commission. (2021). *Commission Decision on the approval on behalf of the European Union of the EU-CELAC 2021–2023 Strategic Roadmap for the implementation of the Brussels Declaration and EU-CELAC Action Plan on Science, Technology and Innovation.* Retrieved May 12, 2022, from https://ec.europa.eu/info/sites/default/files/research_and_innovation/strategy_on_research_and_innovation/documents/eu-celac_strategic-roadmap-2021-2023.pdf

European Climate Law. (2021). *Regulation (EU) 2021/1119 of the European Parliament and of the Council of 30 June 2021 establishing the framework for achieving climate neutrality and amending Regulations (EC) N 401/2009 and (EU) 2018/1999* ("European Climate Law"). Retrieved May 12, 2022, from https://eur-lex.europa.eu/legal-content/EN/TXT/?uri=CELEX:32021R1119

European External Action Service. (2022a). *The role of science in the EEAS.* Strategic communication. Retrieved May 12, 2022, from https://www.eeas.europa.eu/eeas/role-science-eeas_en

European External Action Service. (2022b). *Stronger EU foreign policy via the art of science diplomacy.* Retrieved May 12, 2022, from https://ec.europa.eu/research-and-innovation/en/projects/success-stories/all/stronger-eu-foreign-policy-art-science-diplomacy

European Parliament. (2000). *Lisbon European council 23 and 24 March 2000.* Presidency conclusion. Retrieved August 18, 2022, from https://www.europarl.europa.eu/summits/lis1_en.htm

European Parliament. (2022a). *China-Russia relations: A quantum leap?* Retrieved August 18, 2022, from https://www.europarl.europa.eu/RegData/etudes/BRIE/2022/729349/EPRS_BRI(2022)729349_EN.pdf

European Parliament. (2022b). *Delegations. European Parliament.* Euro-Latin American Parliamentary in Argentina from 11–14 April 2022. Retrieved April 28, 2022, from https://www.europarl.europa.eu/delegations/en/euro-latin-american-parliamentary-assemb/

product-details/20220404DAL10621#:~:text=14%20April%202022-,Euro%2DLatin%20American%20Parliamentary%20Assembly%20in,from%2011%2D14%20April%202022&text=The%2014th%20EuroLat%20plenary%20session,from%2011%2D14%20April%202022
European Union. (2022a). *European Union. External Action. The Diplomatic Service of the European Union.* Science Diplomacy. Strategic Communications, 17.01.22. Retrieved August 15, 2022, from https://www.eeas.europa.eu/eeas/science-diplomacy_en
European Union. (2022b). *European Union External Action. The Diplomatic Service of the European Union.* Latin America and the Caribbean. Retrieved July 18, 2022, from https://www.eeas.europa.eu/eeas/latin-america-and-caribbean_en#:~:text=The%20European%20Union%20(EU)%20and,partnership%20is%20of%20geopolitical%20importance
European Union External Action. (2022). *Foreign affairs council: Press remarks by high representative Josep Borrell after the meeting.* European External Action Service Press Team. The Diplomatic Service of the European Union. Retrieved August 15, 2022, from https://www.eeas.europa.eu/eeas/foreign-affairs-council-press-remarks-high-representative-josep-borrell-after-meeting-0_en
European Union Science Diplomacy. (2022). *European Science Diplomacy Online Course updated.* Retrieved May 12, 2022, from https://www.science-diplomacy.eu/european-science-diplomacy-online-course-updated/
Franc, S., & Karadzija, D. (2019). Quintuple helix approach: The case of the European Union. *Notitia, Journal foe economic, business and social issues,* 5, 91–100. Retrieved August 15, 2022, from https://www.researchgate.net/publication/338475799_Quintuple_helix_approach_The_case_of_the_European_Union
Freeman, C. (1987). *Technology policy and economic performance: Lessons from Japan.* Pinter.
Federal Department of Foreign Affairs of Switzerland Confederation. (2021). *Eleven norms of responsible state behavior in cyberspace.* Retrieved May 12, 2022, from https://www.eda.admin.ch/eda/en/fdfa/fdfa/aktuell/newsuebersicht/2021/04/uno-cyber-normen.html
Focus. (2022). *Giving focus to the Cultural, Scientific and Social Dimension of EU-CELAC relations.* Retrieved May 12, 2022, from https://eulac-focus.net/about-eu-lac/common-vision-based-on-cultural-scientific-and-social-commonalities/u-celac-scientific-cooperation/
Government of Belize. (2021). OAS sixth meeting of ministers and high authorities of science and technology. Belmopan. 8 December 2021. Retrieved August 21, 2022, from https://www.pressoffice.gov.bz/oas-sixth-meeting-of-ministers-and-high-authorities-of-science-and-technology/
Inter-American Development Bank. (2017). I*NTrade. Information system on integration and trade.* Retrieved May 12, 2022, from www.intradebid.org
INTAL. (2016). *Integration & Trade 40.* "Made in Chi-Lat". Retrieved May 12, 2022, from https://publications.iadb.org
JIRI. (2020). Joint press communique of the EU-CELAC JIRI 8 Senior Officials Meeting on Science and Technology. November 2, 2020. Retrieved May 12, 2022, from https://resinfra--eulac.eu/news/joint-press-communique-of-the-eu-celac-jiri-eighth-senior-officials-meeting-on-science-and-technology/
Joint Statement of the Russian Federation and the People's Republic of China on the International Relations entering a New Era and the Global Sustainable Development [Effective]. (2022). Retrieved August 22, 2022, from http://www.lawinfochina.com/display.aspx?id=8215&lib=tax&SearchKeyword=&SearchCKeyword
Kline, S. J., & Rosenberg, N. (1986). An overview of innovation. In R. Landau & N. Rosenberg (Eds.), *The positive sum strategy* (pp. 275–305). National Academy Press.
Leydesdorff, L. (2005). The triple helix model and the study of knowledge-based innovation systems. *International Journal of Contemporary Sociology, 42*, 1–16.
Langenhove, L. V. (2017). *Tools for and EU science diplomacy.* European Commission.
Leijten, J. (2017). Exploring the future of innovation diplomacy. *European Journal of Futures Research, 5*(20), 1–13.

Leydesdorff, L. (2012). The triple helix, quadruple helix, …, and a N-tuple of helices: Explanatory models for analyzing the knowledge-based economy? *Journal of the Knowledge Economy, 3*, 25–35.

Lundvall, B.-A. (Ed.). (1992). National systems of innovation. In *Towards a theory of innovation and interactive learning*. Pinter Publisher.

OAS. (2022). Executive secretariat for integral development. Department of Economic Development. Networks. Retrieved August 21, 2022, from https://www.oas.org/en/sedi/desd/cit/net.asp

Madrid Declaration. (2010). Council of the European Union (2010). *VI European Union-Latin America and Caribbean Summit "Towards a new stage in the bi-regional partnership: innovation and technology for sustainable development and social inclusion"*. Madrid 18 May 2010. Retrieved May 12, 2022, from https://www.europarl.europa.eu/cmsdata/124265/vi_18_5_2010_madrid_en.pdf

McMurtry, A. (2022). *EU plans diplomatic offensive in Latin America to counter Russian, Chinese influence*. AA World. Retrieved August 18, 2022, from https://www.aa.com.tr/en/world/eu-plans-diplomatic-offensive-in-latin-america-to-counter-russian-chinese-influence/2663942#

Meissner, D., & Kotsemir, M. (2016). Conceptualizing the innovation process towards the active innovation paradigm-trends and outlook. *Journal of Innovation and Entrepreneurship, 5*, 14.

Ministry of Foreign Affairs, International Trade and Worship of Argentina (2022). *Argentina busca potenciar el diálogo entre la CELAC y la Unión Europea*. Ministerio de Relaciones Exteriores, Comercio Internacional y Culto. Retrieved July 7, 2022, from https://www.cancilleria.gob.ar/es/actualidad/noticias/argentina-busca-potenciar-el-dialogo-entre-la-celac-y-la-union-europea

Nelson, R. P. (Ed.). (1993). *National systems of innovation: A comparative study*. Oxford University Press.

Ranga, M. (2020). The European Commission's science and knowledge service. *Joint Research Centre*. Brussels: European Commission. Retrieved August 20, 2022, from https://s3platform.jrc.ec.europa.eu/documents/20125/259374/From%20linear%20models%20to%20the%20Triple%20Quadruple%20Quintuple%20N%20tuple.pdf/91b2bd47-447b-5c8d-8c2e-e8873d1ad480?version=1.1&t=1619519846790

Rogers, E. M. (1962). *Diffusion of innovations*. Free Press of Glencoe.

Rothwell, R., & Zegveld, W. (1985). *Reindustrialization and technology*. Longman.

Ruiz, E., & Sanahuja, J. A. (2019). *La Unión Europea y la Agenda 2030 en América Latina: políticas de cooperación en una región de "desarrollo en transición"*. Documentos de Trabajo 8/2019. Madrid. Fundación Carolina. Retrieved August 21, 2022, from https://www.fundacioncarolina.es/wp-content/uploads/2019/06/DT_FC_08.pdf

Sanchez, B. (2018). *New horizons shaping science, technology and innovation diplomacy: The case of Latin American and the Caribbean and the European Union*. Working paper. Bruges: Institute for European Studies. Issue 2018/20. August. p. 10. Retrieved May 12, 2022, from https://5ec1837c-88ac-4ca1-b478-8bfae7f7f027.filesusr.com/ugd/7dd3ca_d9a1a005d-b264b6c915925b049e317a8.pdf

SEGIB. (2022). *Andrés Allamand: "Necesitamos contar con plan de acción para nuestra Diplomacia Científica Iberoamericana"*. Secretaria General Iberoamericana. Retrieved August 21, 2022, from https://www.segib.org/andres-allamand-necesitamos-contar-con-un-plan-de-accion-para-nuestra-diplomacia-cientifica-iberoamericana/

Selleslaghs, J. (2017). EU-Latin American science diplomacy. *Working paper EL-CSID, Institute for European Studies*, issue 2017/8. September. Retrieved May 12, 2022, from https://5ec1837c-88ac-4ca1-b478-8bfae7f7f027.filesusr.com/ugd/0bc3be_aef41ff5b1e7451bb7d9063a0c7075d0.pdf

The State Council Information Office of the People's Republic China. (2021). *China: Democracy that works*. Retrieved August 20, 2022, from http://english.scio.gov.cn/whitepapers/2021-12/04/content_77908921.htm

Torres Jarrín, M. (2021). Indo-Pacific region: A political concept or a geopolitical reality? In J. R. Argumosa Pila, A. Garofali Acosta, & M. Torres Jarrín (Eds.), *The future of the Asia-Pacific in the New World order. Emerging markets & emerging powers* (p. 26). Tirant Humanidades.

Torres Jarrín, M. (2018). EU-CELAC: A multiplayer interregionalism. Redefining the Atlantic area. In G. L. Gardini, S. Koschut, & A. Falke (Eds.), *Interregionalism and the Americas*. Lexington Book.

Yawson, R. M. (2009). *The ecological system of innovation: A new architectural framework for a functional evidence-based platform for science and innovation policy*. Paper presented at the Future of Innovation Proceedings of the XXIV ISPIM 2009 Conference. Vienna Munich Personal RePEc Archive.

# Chapter 9
# Cyberdiplomacy and Techplomacy in EU–CELAC Relations

**Abstract** This chapter will return to the themes of Cyberdiplomacy and techplomacy and the scope for collaboration between EU and CELAC. It will look at the importance of both in the development of the digital economy, and relate it to existing EU–CELAC joint programmes. It will suggest ways in which CELAC and the EU could collaborate, both in promoting a ruled-based cyberspace and in strengthening digital resilience. It will also explore the scope for inter-regional collaboration in these areas in relation to managing relations with internet giants and generating local competitors.

**Keywords** Cyberdiplomacy · Techplomacy · EU · CELAC · Cyberspace norms · Cybersecurity · Inter-regionalism 4.0 · Global governance 4.0

## 9.1 Cyberdiplomacy and the Scope for Collaboration Between EU–LAC Relations

During the last 20 years international cybersecurity has been a major issue on the United Nations' agenda and several working groups have been set up to promote exchanges of experience and dialogue among experts and governmental representatives on the potential threats to peace and international security posed by the use of information and communication technologies (Federal Department of Foreign Affairs, 2021).

Between 2004 and 2021 the United Nations has led a cybersecurity dialogue process through the work of the United Nations Group of Governmental Experts (UN-GGE). Six Groups of Governmental Experts (GGE) on cybersecurity have been convened. The current process is the United Nations Open-Ended Working Group (OEWG) on Developments in the Field of Information and Telecommunications in the Context of International Security set up by the UNGA. The OEWG is into its second mandate 2021–2025.



M. Torres Jarrín, S. Riordan, *Science Diplomacy, Cyberdiplomacy and Techplomacy in EU-LAC Relations*, United Nations University Series on Regionalism 25, https://doi.org/10.1007/978-3-031-36868-4_9

Cyberspace is an intangible territory that encompasses all the tangible territories around the world (countries). Companies exist whose actions have geopolitical repercussions for governments. In 2016 at the NATO Summit, held in Warsaw 8–9 July, its Member States recognized that cyberspace was a new domain of operations, just like the land, air, sea and space. During the past decades, malicious cyber activity has increased, from ransomware to malware motivated by espionage or political objectives.

In this new domain, operations are carried out by natural persons or legal entities (private and public). Their activities may be legal or illegal. Goods and services are offered that can be promoted by people or machines. Artificial Intelligence through the management of Big Data processes, systematises and analyzes the data that navigates in cyberspace. This data can be used by government agencies to prevent crimes and conflicts, or they can also design new foreign strategies with allied countries or enemy countries. Public agencies or private businesses can use cyberspace as a way to destabilize regimes, whether democratic or authoritarian. The data analysed by companies that utilise AI is also used to create new products and improve marketing strategies for goods and services.

In the last few years, cybersecurity has become an important issue on the global agenda and has become a priority issue in countries' foreign and security policy strategies. At the global level we can observe five international and regional organisations which deserve particular attention: the UN Group of Government Experts (UN-GGE), the Organization for Security Cooperation in Europe (OSCE), ASEAN Regional Forum (ARF), the Organization of American States (OAS) and the European Union (EU).

Regulating cyberspace requires enhanced collaboration between stakeholders and across sectors. The Latin American and Caribbean countries have not discussed or reached agreements as a region, and much less within the CELAC (Community of Latin American and Caribbean States) framework, despite this being the latest regional integration process and highest political forum at the regional and inter-regional level. CELAC is the official counterpart that leads the bi-regional negotiations between EU and LAC countries through the EU–CELAC Summits system. Curiously, Latin American and Caribbean countries created CELAC because they did not feel represented in the Organization of American States (OAS), but it is the OAS framework where dialogues are developed on how to face the challenges that cyberspace represents for countries. In 2015, the OAS was the first organisation in the world to embark on a deep and broad study of cybersecurity capacities across a whole region, assessing the status of developments in Latin America and the Caribbean. Twelve LAC countries have a National Cybersecurity Strategy, 6 countries are developing a National Cybersecurity Strategy and just 8 of 33 CELAC Member States are part of the Convention on Cybercrime, otherwise known as the Budapest Convention.

It should be noted that the Budapest Convention is the first international treaty seeking to address Internet and cybercrime by harmonising national laws and increasing cooperation among countries (Budapest Convention, 2001). The convention provides for the criminalisation of conduct—ranging from illegal access, data and systems interference to computer-related fraud and child pornography—supports procedural powers to investigate cybercrime and secure electronic evidence in relation to any crime; and promotes efficient international cooperation (Council of Europe, 2021).

We can also point to the leading role of the London Process, launched by the former Foreign Secretary of the UK, William Hague in 2011. The London Process promoted a series of informal international meetings whose aim was to generate a consensus on responsible behaviour in cyberspace and established the Global Forum on Cyber Expertise (GFCE), where the OAS (OAS and IDB, 2016) and European Union were founding members. Some European, Latin American and Caribbean countries are state members: Argentina, Austria, Belgium, Chile, Estonia, Finland, France, Germany, Hungary, Latvia, Mexico, The Netherlands, Norway, Peru, Romania, Spain, Suriname, Sweden, Switzerland and the UK (Delhi Communiqué on a GFCE Global Agenda for Cyber Capacity Building, 2017).

The GFCE was established during the 2015 Global Conference on Cyber Space held in The Hague. It was launched by the Dutch Government along with 41 ministers and other high-level representatives from business and international organisations, who recognised that societies are becoming increasingly digitised, interconnected and dependent on the cyber domain for communication, innovation and sustainable social development and economic growth and for these reasons they created a pragmatic, action-oriented and flexible forum that will complement and reinforce existing bilateral, multilateral, multiparty, regional and international efforts to build cyber capacity and expertise and avoid duplication and overlap (The Hague Declaration on the GFCE, 2015).

The activities of the GFCE will have a geographic or thematic focus on cybersecurity, cybercrime, data protection and E-governance (GFCE, 2015). In 2018, the GFCE Working Groups were established, following the Delhi Communiqué on a Global Agenda for Cyber Capacity Building:

- Cybersecurity policy and strategy
- Cyber incident management and critical information infrastructure protection
- Cybercrime
- Cybersecurity culture and skills
- Cyber Internet standards

In the Western Hemisphere on 8 June 2004, the OAS General Assembly adopted at the fourth plenary session the "Adoption of a comprehensive Inter-American Strategy to combat threats to Cybersecurity: A multidimensional and multidisciplinary approach to creating a culture of cybersecurity". This document was the result of the OAS Conference on Cybersecurity and the Inter-American Committee against Terrorism (CICTE).

The OAS Conference on Cybersecurity, held in Buenos Aires on 28 and 29 July 2003, recognised the gravity of cybersecurity threats to the security of critical information systems, critical information structures and economies throughout the world. The Inter-American Committee against Terrorism (CICTE) at its fourth regular session considered the document "Framework for Establishing an

Inter-American CSIRT Watch and Warning Network" and decided to hold a meeting of government cybersecurity experts, in March 2004 in Ottawa, to prepare the recommendations for the adoption of a comprehensive Inter-American Strategy above mentioned.

The Inter-American Strategy further recognised that an effective framework was needed for protecting the networks and information systems that constitute the Internet; furnishing users and operators of the Internet with information to help them secure their computers and networks against threats and vulnerabilities; fostering public–private partnerships with the goal of increasing education and awareness and working with the private sector, which owns and operates most of the information infrastructures on which nations depend; and fostering the adoption of cybercrime policies and legislation that will protect Internet users and prevent and deter criminal misuse of computers and computers networks. At the same time, OAS Member States are committed to fostering a culture of cybersecurity that deters misuse of the Internet and related information systems and encourages the development of trustworthy and reliable information networks.

The CICTE has a Cybersecurity Program whose objectives are to:

- Support OAS's Member States in building technical and policy capacities to successfully prevent, identify, respond to and recover from incidents.
- Enhance robust, effective and timely information sharing, cooperation and coordination among cybersecurity stakeholders at the national, regional and international levels.
- Boost access to knowledge and information on cyber threats and risks by public, private and civil society stakeholders as well as Internet users.

Among its achievements the programme has the following results:

- Design and implementation of 17 national cybersecurity strategies developed in Latin America and the Caribbean region.
- Two regional reports focused on the cybersecurity capacity of OAS Member States.
- Over 2000 women trained in cyber and more than 600 young students in digital security and cybersecurity.
- Capacity programmes for 15,000 citizens and public and private officials trained in cyber operations, cybersecurity, diplomacy, cybersecurity leadership and international cybersecurity regulations.
- More than 20 publications focused on the regional level: critical infrastructure, cybersecurity education and workforce development, digital security, online gender violence, data classification, among others.

This programme organises Cybersecurity Symposiums with discussion topics related to cybersecurity trends, data analysis, Internet privacy and security trends, public cybersecurity policies, cybercrime etc. The last symposium, held in September 2021, included such topics as: the role and future of cybersecurity in the post-COVID-19 transition, ransomware trends, women leaders in incident response teams, new rules for data privacy, cybersecurity and privacy trends in social media and innovation and emerging technologies (CICTE, 2021).

Cybersecurity is included in EU–CELAC relations. In the EU–CELAC 2021–2023 Strategic Roadmap for the implementation of the Brussels Declaration and EU-CELAC Action Plan on Science, Technology and Innovation, the countries considered that cybersecurity is a vital enabler of knowledge, productivity and growth in this digital age (European Commission, 2021). Taking account of all the variables mentioned above, the EU has created the BELLA (Building the Europe Link to Latin America) programme, which is divided into two projects: BELLA-S and BELLA-T. The first is a direct submarine cable linking the two regions. The cable will provide for the transatlantic data-sharing needs of the European and Latin American research and education communities for the next quarter of a century and it is expected that BELLA-S will implement sufficient capacity between the GÉANT (the Pan-European data network for the research and education community) and RedCLARA (Latin American Advanced Networks Cooperation) networks. The second project, BELLA-T is the terrestrial optical fibre network infrastructure for RedCLARA.

Since 2004 RedCLARA has provided regional interconnection and connection with the rest of the world via GÉANT, via US Network Internet2 via Africa's advanced networks as UbuntuNet Alliance, WACREB, ASREN and Asia networks such as APAN, TEIN, CAREN and in Oceania via AARNET. With the BELLA programme Latin American countries are reducing their dependency on the infrastructure of the USA. Latin American countries explored also the potential Trans-Pacific Internet cable proposed by former Chilean President Bachelet at the CELAC Summit in 2018.

Most Latin American countries voted for the establishment of the Open-Ended Working Group (OEWG) at the United Nations to debate cooperation on stability in cyberspace. This dismayed the USA, who saw the Russia-sponsored OEWG proposal as competing with its proposal for the creation of a new UN Group of Governmental Experts (Van Raemdonck, 2020). In the EU, in September 2017, the European Commission proposed a Cybersecurity Act as part of a wide-ranging set of measures to deal with cyberattacks and to build strong cybersecurity in the EU. After that the European Parliament, the Council and the European Commission reached a political agreement on 10 December 2018. On 12 March 2019, the European Parliament Plenary in Strasbourg adopted the European Cybersecurity Act (European Parliament and Council, 2019) with 586 votes to 44 against and 36 abstentions (European Parliament, 2018). The objective of the EU Cybersecurity Act is to achieve a high level of cybersecurity, cyber resilience and trust in the European Union by setting objectives, tasks and organisational matters for a strengthened and renamed European Union Agency for Cybersecurity (ENISA), with a new mandate; and the creation of a framework for voluntary European cybersecurity certification schemes for Information and Communications Technology (ICT) products, services and processes (The EU Cybersecurity Act, 2019).

On 16 December 2020, the European Commission and the High Representative of the Union for Foreign Affairs and Security Policy presented a new EU Cybersecurity Strategy to bolster Europe's collective resilience against cyber threats and help ensure that all citizens and businesses can fully benefit from trustworthy and reliable services and digital tools. The strategy also allows the EU to take the lead on international norms and standards in cyberspace, and to strengthen cooperation with partners around the world to promote a global open, stable and secure cyberspace, grounded in the rule of law, human rights, fundamental freedoms and democratic values (European Commission, 2020).

The EU and LAC countries have as common goals an open, stable and secure cyberspace as well as rules and human rights. On 2 December 2021, leaders from the EU-LAC countries launched a Digital Alliance to advance convergence in rules and standards and promote digital innovation (European Council, 2021). At the same time, they plan to develop a human-centred digital transformation plan by 2022 with the objectives of further aligning policies, consolidating digital connectivity and facilitating business partnerships for innovative digital solutions (European External Action Service, 2021).

In June 2005, the LAC countries approved the Action Plan for the Information Society. The agenda sought to be a catalyst for regional cooperation on digital matters and a mechanism to promote policy design, capacity-building and political dialogue on the challenges and opportunities that digital transformation creates for Latin American and Caribbean societies and includes 9 areas of action and identifies 39 specific goals for implementation (ECLAC, 2021). Under these 9 areas the EU–CELAC countries can create a new common agenda for bi-regional cooperation:

1. Digital infrastructure
2. Digital transformation and digital economy
3. Digital government
4. Inclusion and digital skills and other competencies
5. Emerging technologies for sustainable development
6. Trust and digital security
7. Regional digital market
8. Digital regional cooperation
9. Combating the pandemic and facilitating economic recovery and reactivation

In particular, the areas number 7 and 8 specify in goals 27, 28, 29, 30 and 31:

- Goal 27: Promote a regional digital market strategy, including in the framework of regional sub-regional integration mechanisms, that facilitates cross-border e-commerce and digital trade through integration of digital infrastructure, regulatory harmonisation, free flow of data with trust, in accordance with domestic legislation, trade facilitation, improved postal and logistic services and regulatory frameworks that encourage innovation in digital payment services.
- Goal 28: Facilitate greater regional coordination through a digital integration plan that establishes a common vision and goals, with mechanisms for dialogues and coordination with existing regional and sub-regional organisations.
- Goal 29: Promote strengthened regional digital cooperation in Internet governance processes by reinforcing national, regional and sub-regional multistakeholder dialogue mechanisms.

- Goal 30: Strengthen the institutional structure of the entities responsible for designing, implementing and following up on digital agendas. Formalise and coordinate multistakeholder participation in the development of such policies.
- Goal 31: Promote digital regulatory coherence at the regional level, especially on data protection, cross-border data flows, cybersecurity, e-commerce and digital trade, consumer protection and rights on online platforms, and interoperability between digital signatures and digital identity systems in the region, in line with domestic policy and regulatory frameworks.

## 9.2 Techplomacy Cooperation Between EU and CELAC: Towards an Inter-regionalism 4.0

On the one hand, from a defence perspective, in the case of regulation of cyberspace, we find the American countries (North American, Latin American and Caribbean) more advanced than the European ones. On the other hand, in terms of regulation of digital services, we find the European countries more advanced than their American partners. This is because there is a common market—the EU Common Market—unlike the situation in the Americas.

In March 2016, the Inter-American Development Bank (IADB) and the Organization of American States (OAS) published the first edition of the report "Cybersecurity: Are We ready in Latin America and the Caribbean?" The purpose was to provide a picture of the state of cybersecurity and guidance about the next steps that should be pursued to strengthen national cybersecurity capacities in Latin American and the Caribbean countries. The report indicates that every year, millions of new LAC users connect to the Internet for the first time and the rise in the number of cyberattacks has spurred an increase in the interest in cybersecurity in the region.

In June 2016, the EU published its "Global Strategy for the European Union's Foreign and Security Policy" that included cyber in the priorities, building also on the approach of the Digital Single Market. One year later, the EU published a Joint Communication of the European Parliament and the Council: "Resilience, Deterrence and Defense: Building strong cybersecurity for the EU" 2017) that mentioned that tens of billions of "Internet of Things" devices that were expected to be connected to Internet by 2020. It expressed concern that cybersecurity is not yet prioritised in their design and the failure to protect the devices which will control power grids, cars, transport networks, factories, finances, hospitals and homes could have devastating consequences and cause damage to consumer trust in emerging technologies, besides the risks of politically motivated attacks on civilian and military targets (European Commission, 2017).

In 2017, the Ministry of Foreign Affairs of Denmark presented its Foreign and Security Policy Strategy 2017–2018, which included three new concepts that marked a new stage in the history of international relations: Techplomacy, Tech

Ambassador and Tech Embassies (Ministry of Foreign Affairs of Denmark, 2017). The strategy highlighted the rapid development of areas such as artificial intelligence, the Internet of things, cybersecurity, machine learning, among others; generated by the Fourth Industrial Revolution (Schwab, 2015), which was transforming all sectors of society and generating an impact both in national markets as well as in the global balance of power and established values. As a consequence, the external actions of the Big Tech Companies were affecting key issues in foreign policy (Torres Jarrín & Riordan, 2019). According to Schwab "the technology revolution will change the way we live, work and interact. In scale and complexity, the transformation will be unlike anything mankind has ever experienced". He predicted a convergence between the physical, digital and biological worlds that would impact governments and global governance (Schwab, 2016). The Danish Foreign Minister, Jeppe Kofod opined that: "We've been too naïve for too long about the tech revolution. We need to make sure that democratic governments set the boundaries for the tech industry—and not the other way around. That's where the Danish Techplomacy initiative comes in" (Kofod, 2021).

On 21 May 2018, the Inter-American Development Bank organised a conference on "Techplomacy. Reinventing Diplomacy in the Digital Age". The keynote speaker was the world's first TechAmbassador, Casper Klynge from the Danish Government. Ambassador Klynge shared his experiences and explained why Techplomacy is important for countries. He explained that the TechAmbassador has a global mandate and a physical presence across three time zones in Silicon Valley, Copenhagen and Beijing, transcending borders and regions and rethinking diplomacy. Techplomacy offers the Danish government the opportunity to have a direct and frank dialogue with tech companies in order to try and influence the direction of technology, at the same time seeking to influence the international agenda around tech policy in accordance with Danish interests and values, including through new multilateral alliances, and multistakeholder partnerships (Ministry of Foreign Affairs of Denmark, 2022).

In February 2021, the Ministry of Foreign Affairs of Denmark published the new "Denmark's Tech Diplomacy Strategy 2021–2023". The new strategy emphasizes that "the balance of power in tech has tipped. Today, a small number of people in a small number of private companies hold immense sway over the direction of an increasingly global society. Democratic government, parliaments, and the people they legitimately represent have conversely lost some of their ability to steer our democratic societies" At the same time, the strategy mentioned: "there is a more critical view of the tech giants. There is no question that the tech sector must be regulated—the question is how to regulate it" (Ministry of Foreign Affairs of Denmark, 2021).

On the other hand, the strategy indicates a "Digital World Map—Market Valuation of Major Global Tech Companies" that identifies 17 Big Tech Companies of which five are from the USA (Amazon, Microsoft, Apple, Alphabet, Facebook—renamed as Meta), three China (Meituan, Tencent, Alibaba), two Sweden (Ericsson and Spotify) one South Korea (Samsung), one Australia (JB), one France (Dassault), one Germany (SAP), one The Netherlands (ASML), one Argentina (Mercado Libre) and one South Africa (Naspers). The Danish Tech Diplomacy Strategy 2021–2023 focuses on three priorities: Firstly, responsibility (to work to ensure that the tech giants meet their societal responsibility and operate on a level playing field); secondly, democracy (Denmark wants to be the champion of global digital rules and regulations that build on democratic values and human rights) and thirdly, security (technology to support Denmark's safety and security).

According to the World Economic Forum in its Innovation Geographies 2022 report, of the top 25 global cities for innovation, 10 are in North America (San Jose, San Francisco, Boston, New York, Seattle, Los Angeles, San Diego, Austin, Chicago and Raleigh-Durham), 6 in Europe (Paris, London, Berlin, Stockholm, Munich and Helsinki), 7 in the Asia-Pacific region (3 Chinese (Beijing, Shanghai and Shenzhen), 2 Japanese (Tokyo and Osaka), Seoul and Singapore) and 1 in the Mediterranean (Tel Aviv) (Bateman, 2022). If we analyse this from a regional perspective, we can observe most Tech-Cities are concentrated in North America, Europe and the Asia-Pacific regions, respectively, and there is only one tech company in each of Africa and Latin America. In this sense, we need to create a Tech Diplomacy Inter-regionalism. The EU–CELAC strategic partnership can promote this new inter-regionalism. Global governance needs to incorporate the voices of the Big Tech Companies which have led the great transformation of our societies. We need an Inter-regionalism 4.0 for a Global Governance 4.0 and to achieve this it is essential to have a Tech Diplomacy Policy in each region and then develop a Tech Diplomacy Inter-regionalism.

To design and implement Tech Diplomacy Inter-regionalism, the first step is creating a Tech Diplomacy at the regional level. Prior to that, it is necessary to understand: what is Tech Diplomacy? Even more important is that the foreign action services of the countries understand it and develop a foreign policy on Tech Diplomacy.

In Europe, France, Germany, Switzerland and Estonia have followed in the footsteps of the Danish government. Four other EU Member States have developed a foreign policy with a focus on Tech diplomacy: Denmark, France, Germany and Estonia. In December 2017, the French government launched its "France's International Digital Strategy" and later in 2019 appointed Ambassador David Martinon as "Ambassador for Digital Affairs" in Silicon Valley. The French strategy focuses on five areas (Ministère de L'Europe et des Affaires Étrangéres, 2017):

1. Promote and supervise the development of innovations and the control of state-of-the-art technologies, particularly those related to Artificial Intelligence.
2. Guarantee the security and international stability of the digital space.
3. Promote human rights, democratic values, and the French language in the digital world.
4. Strengthen the influence and appeal of French digital players.
5. Contribute to Internet Governance.

In August 2018, the Federal Ministry of Foreign Affairs of Germany appointed Ambassador Hinrich Thoelken as "Special Representative for International Digitalization Policy and Digital Transformation". His mission consists in analysing the effects of digital transformation in the economy, society and at the international level in order to develop a foreign policy strategy adapted to the digital age (Federal Foreign Office of Germany, 2019). According to Ambassador Thoelken: "The German Federal Foreign Office is directly affected by the rapid advance of technological changes [...]. This applies to his mission as a foreign policy agency and service provider, as a government ministry and employer. Technologies such Big Data, Machine Learning and Quantum Computing can lead to new global balances of economic and political power. [...] Non-state actors are increasingly able to set the agenda. Technological progress also raises the question of how we will continue to fulfill our sovereign tasks in the future" (Thoelken, 2019).

The Swiss federal government appointed Dr. Joan Fanzun as "Special Envoy for Cyber Foreign and Security Policy". In November 2020, the Swiss government presented its "Digital Foreign Policy Strategy 2021–2024", which aims to start a new phase by shaping the governance of digital problems. The strategy responded to a proposal that the Federal Council called "Postulate 17.3789" which was proposed on 28th September 2017 and whose objective was to elucidate "How Switzerland could become the world epicenter of international governance of cyberspace". The postulate also contemplated the creation of a "Geneva Convention for Digitalization" and foundation of a neutral organisation in Geneva that is in charge of ensuring compliance with the convention (Federal Council of Switzerland, 2020). It should be noted that Switzerland has always been a country that has exercised leadership regarding the creation of norms in cyberspace and regulation of the Internet. In 2003, Geneva hosted the Internet Governance Forum, and it was a pioneer in proposing confidence-building measures in the Organization for Security and Cooperation in Europe At the same time, it was the chairman of the Open-Ended Working Group on Cybersecurity of the United Nations. Along with Singapore, Estonia, Rwanda and the Netherlands, Switzerland has been one of the most prominent countries in promoting more inclusive digital governance globally (Kurbalija, 2021).

In 2018, the Estonian government appointed Heli Tiirmaa-Klaar as "Ambassador at large for Cyber Diplomacy" and, in 2019 she was appointed as the director of the Department for Cyber Diplomacy. For Ambassador Tiirmaa-Klaar: "It is important to continue the efforts that already have started in Estonia on analyzing the applicability of existing international law in cyberspace, as well as to support international processes promoting cyber norms and confidence-building measures. Estonia will also continue to actively participate in ongoing EU and NATO cyber initiatives and to strengthen bilateral cooperation in the field of cyber security with its allies" (Ministry of Foreign Affairs of the Republic of Estonia, 2018). During the period from 2012 to 2018, Ambassador Heli Tiirmaa-Klaar was head of policy coordination for cyberspace at the European External Action Service. She was in charge of designing and leading a European Strategy for Cyberspace Security. Between 2013 and 2017, she was responsible for both the development of EU cybersecurity strategies and the implementation of the EU Cyber Defense Policy Framework.

In the case of Latin American and the Caribbean countries, there is still no ministry of foreign affairs that is working on strategy or foreign policy focused on Tech

Diplomacy. However, Latin American and Caribbean governments are working on how to regulate cyberspace. Some of them already have a strategy in place, such as Colombia, Jamaica, Panama and Trinidad and Tobago. Others are in the process of developing one, such as Costa Rica, Dominica, Peru, Paraguay and Suriname (Maciel et al., 2016).

In 2018, in the Bay Area (that includes Silicon Valley) diplomatic missions aiming to interact with the Big Tech Companies included (Horejsova et al., 2018):

- Embassy: Tech Embassy of Denmark
- Consular representation: Germany, India and Greece
- Consular representation and an innovation centre: Austria, Switzerland and The Netherlands
- A state investment promotion agency: Czech Republic
- Honorary Consulates: Hungary and Finland
- From the embassies in Washington: Tunisia, Turkey and Nigeria

## 9.3 Projects and Programmes in Cyber Diplomacy and Tech Diplomacy

Digital technologies are transforming markets, societies, governments, international relations and global governance. The EU and CELAC can incorporate in their bi-regional dialogue the exchange of ideas on cyberspace regulation policies through the development of a tech diplomacy policy. Within the EU–CELAC summits framework, the countries can promote the creation of a forum that brings together stakeholders from Big Tech Companies, Governments, Academia, Think Tanks and Civil Society, where they will discuss what type of regulation is necessary.

Consensus can be a good tool to develop some policies, but more important is to advance a dialogue that does not exist yet. Our society lives in a global, digital and physically and virtually interconnected world, for which reason countries need to create international norms. Inter-regionalism can be the better mechanism to implement a Tech Diplomacy Policy. But it is necessary to create an inter-regionalism 4.0 to respond to the challenges of the Fourth Industrial Revolution. We need to rethink the global governance model and reconsider the actors: the Nation States, the regional integration organisations or the international organisations. But they are not the only and exclusive actors in the international arena. The Big Tech Companies are the new geopolitical actors, and their decisions must be taken into account by the other actors.

In the Latin American and Caribbean framework, there is an interesting experiment led by the Permanent Secretariat of the Latin American and Caribbean Economic System (known by its acronym in Spanish as SELA) and the European Institute of International Studies (EIIS) under the mandate of the XLVI Latin American Council on promoting the development in Latin America and the Caribbean of negotiating capacities for the governance of cyberspace. Both institutions implemented two capacity-building programmes for diplomats and other government officials. The first edition was a Specialization Course on Cyber Diplomacy aimed at describing the application of diplomacy to the political and geopolitical problems that have arisen in cyberspace. The second edition was a Specialisation course on Techplomacy with the purpose of describing the interaction between States and large companies in the technology sector (SELA, 2022). The first edition was organized by SELA and EIIS in 2021 and, the second was organized by SELA and EIIS with the collaboration of the Institute of European Studies and Human Rights at the Pontifical University of Salamanca (Spain) and held in 2022. In both editions, 362 diplomats and other governments participated in the cyberdiplomacy course and 329 participants in the Techplomacy course. This bi-regional cooperation was a pioneer of its kind. These courses could be replicated in the EU–CELAC cooperation framework, between the European External Action Service (EEAS) and the Latin American and Caribbean Foreign Services under the umbrella of SELA.

In August 2022, the European Union announced that it will open a new liaison in Silicon Valley to Big Tech Companies, in other words a "Tech Embassy", giving European regulators direct access to major digital companies and vice versa. The office opened on 1 September 2022 and comes as the EU prepares to implement two landmark technology regulations: the Digital Markets Act and the Digital Services Act. (World Economic Forum, 2022). According to Charles Manoury: "The office will strengthen the EU's capacity to reach out to key public and private stakeholders, including policy-makers, the business sector and civil society in the digital technology sector". The office operates under the authority of the EU delegation in Washington DC and will be led by Gerard de Graaf, director of digital economy in the European Commission (The European Sting, 2022). The EEAS-SELA can develop cooperation in areas such as cyberdiplomacy and techplomacy that can contribute to creating a bi-regional agenda that can be implemented in its external action with Big Tech Companies. And the leadership in the regulation of cyberspace can be a common external action between EU-CELAC.

The OAS "CICTE's Cybersecurity program" which aims to build technical and policy-level cybersecurity capacities is a promising experience about which cooperation between American and European governments can be constructed. The EU-United State Trade and Technology Council serves as a forum to coordinate approaches to key global trade, economic and technology issues, as well as to deepen transatlantic relations through common values (European Commission, 2022a). Among the objectives of the EU-US Trade and Technology Council are to strengthen the technological and industrial leadership of both sides. In the Second Ministerial-Level Meeting of the EU-US Trade and Technology Council, held in Paris on 15 and 16 May 2022, the countries decided the following points with respect to emerging technologies (European Commission, 2022b):

- To establish a Strategic Standardisation Information mechanism to promote and defend common interests in international standardisation activities.

- To develop shared methodologies for measuring Artificial Information trustworthiness and risks.
- To develop a roadmap for future communication technologies beyond 5G and 6G.
- To strengthen cooperation on key aspects of platform governance, such as content moderation, algorithmic amplification, data access for research and the protection of minors.

Given that the transatlantic area is built on three regions—North America (Canada and the USA), Latin America and Caribbean and European Union—a new inter-regionalism, a "Triple Inter-regionalism" could be developed. This "triple inter-regionalism can reinforce the West's position in the world order. The USA cannot lead alone the international system without allies, and what better than having an alliance consisting of the three regions that make up the Atlantic area.

Academia has not taken into consideration two important facts within the EU's foreign policy. Firstly, since 2011, the EU is a permanent observer of the OAS. Secondly, the previous cooperation programme of the EU for LAC region for the 2014–2020 period was divided into two main areas:

1. Activities at continental level
2. Sub-regional programme for Central America

In 2014, for the first time, the EU included within its cooperation with Latin America and Caribbean countries, a cooperation that seeks to promote activities at a continental level instead of doing so at a regional or sub-regional level. This decision made sense, because in previous cooperation programmes the EU had opted for regional integration of Latin America and the Caribbean as a whole region. For that reason, the EU financed the four main sub-regional integration processes: EU-CAN cooperation programme (2002–2006), EU-SICA cooperation programme (2007–2013), EU-CARICOM cooperation programme (2007–2013) and EU-MERCOSUR cooperation programme (2007–2013). After these four cooperation programmes, the EU's only achievement was to sign the EU-SICA agreement. Hence, the 2014–2020 cooperation programme only considered cooperation with Central America, and the EU decided to develop actions at a continental level.

On the other hand, since 2004 China has been a permanent observer in the OAS. In fact, in the last Dialogue of the Heads of Delegation of the Member States, the Secretary General and the Assistant Secretary General with the Permanent Observers, held in Guatemala on 10 November 2021 the Chinese ambassador Qin Gang, permanent observer to the OAS, indicated: "China has invested more than 3.5 million U.S. dollars through the China-OAS Cooperation Fund and Scholarship Project Agreement to nearly 100 projects of the OAS General Secretariat. China also is a member of the Inter-American Development Bank, established the "China-Co-financing Fund for Latin America and the Caribbean" and annually organised the "China-LAC Business Summit" (Gang, 2021).

On the European Union side, Javier Niño Perez, Deputy Director General for the Americas of the European External Action Service indicated that European Union had been the largest contributor to the COVAX global initiative and announced a new cooperation programme endowed with 3 billion euros to continue advancing prosperity and stability in the region. Likewise, he pointed out: "Europe's total commitment to the region, total commitment to the organization. We want to move forward together towards a green, digital, inclusive recovery and development in which Europe and the Americas must walk hand in hand" (Niño, 2021).

Taking into account all these various foreign policy initiatives from the European Union, China, the USA and Latin America and Caribbean countries, the OAS could be an ideal platform for working on issues related to cyberdiplomacy and techdiplomacy.

## 9.4 EU–CELAC Relations and in the New World Order

The twentieth century ended with a process of globalisation (Crafts, 2000), with interdependent economies (Streeten, 2001) through the creation of global values chains (McKinsey Global Institute, 2019), with interconnected societies and with capitalism and liberal democracy as an economic and political model.

The twenty-first century begins with a digitalisation process that is transforming societies, governmental sector, private sector and cultural sector. The irruption of digital technologies has transformed the way of thinking, designing, producing and communicating. The Big Tech Companies are emerging as the new powers in the international arena. The power of the nation state is in question, as are several regional integration processes. The rise of populism, nationalism, and protectionism (Norris & Inglehart, 2019; Rodrik, 2019; Collier, 2018) seems to be a measure of resistance to these processes of change. At the beginning of the twentieth century, the rise of pollical parties and nationalist governments contributed to outbreak of the First World War. Later populist governments pursuing protectionist policies contributed to the outbreak of the Second. These threats are in force today. Just as the establishment of multilateralism was a response to avoid wars and prevent armed conflicts, now the world is experiencing another similar moment. But the context is different. It is necessary to implement regulations in the digital sector (European Commission, 2020b), as well as create new international theories, policies and development models in accordance with the times in which we live.

The pandemic has been an accelerator of change (OCDE, 2020). Changes that many countries have resisted making, wanting to maintain their status quo power structures, defend their political, military and economic privileges. Global governance needs to incorporate new actors that until now have remained on the sidelines of decision-making on the major issues on the international agenda. The core of countries that lead the world order has to be reconfigured because there are emerging powers, some of them are great and historic world powers that have returned to the international scene and claim a place according to their geopolitical power and political and economic influence, whether at the regional or global level.

Inter-regionalism faces a change of era. The Fourth Industrial Revolution has created an Industry 4.0 that demands new international standards. Geographic spaces continue to be important. However, the creation of a new region—cyberspace—must be taken into account. This region has not been created by nation states but by the Big Tech Companies. This cyberspace region has to be incorporated into the foreign policies implemented by governments (at national and regional levels). It is necessary to create an inter-regionalism 4.0 that involves governments, regional integration organisations, international organisations and civil society.

The first half of the twentieth century developed through the implementation of policies decided fundamentally at nation state level. But with the birth of the European integration process, other regional and sub-regional integration processes were created around the world. These regional organisations became international actors that contributed to the design and implementation of international, regional and national development policies. Suddenly the world was no longer divided only into nation states: it was divided also into regions. For its part, the new "cyberspace region" grew in parallel. Progressively, many of the actions of governments, businesses and citizens in the physical world passed into cyberspace, increasing the relevance and influence of cyberspace in the international arena.

Today wars are hybrid. Many major attacks are produced using the network that allows you to navigate through cyberspace. A cyberattack can threaten civilian populations through attacks on energy systems or gas, oil or food supplies and other critical infrastructure. The impact of these attacks can be more widespread than those of conventional missiles or bombs. Attacks on health, electricity or water systems can cause chaos in any city in the world causing economic and social crises. If we add to this the attacks on financial and security and systems, the consequences for a country's social and economic stability can be serious.

As mentioned in previous chapters, EU–CELAC relations had been stagnant since 2015, the year in which the last EU-CELAC Summit of Heads of State and Government was held. The political crisis in Venezuela gave rise to two institutional crises in CELAC, one of an internal nature, which prevented consensus from being reached regarding the political situation in Venezuela. CELAC members were divided into two groups: a group of countries that supported the government of President Maduro and another group that was against the government considering that Venezuela was not a democracy, and therefore did not respect the rule of law and the separation of powers. The other crisis was external, which affected foreign relations with third parties, including relations with the European Union.

But the EU was not experiencing its best moment either. The change of government in the USA, with the election of President Trump, who redefined all of US foreign policy, inevitably had an impact on traditional transatlantic relations. There was a freeze in US relations with its European partners, and discrepancies about the future of transatlantic relations were a constant variable between both parties. The trade war between the USA and China, Brexit, the invasion of Ukraine and the tensions in relations between the EU and Turkey and Iran were all major issues that occupied much of the time of Europeans, resulting in less attention being paid to relations with Latin America and the Caribbean.

The Spanish Presidency of the EU re-launched the series of EU–CELAC Summits in July 2023. The German presidency of the EU Council tried to do so in

December 2020 when it organised an informal meeting of the Ministers of Foreign Affairs of the EU27 and LAC. Although the countries expressed their will to return to the bi-regional political dialogue at the highest level, that is at the head of state and government level, nothing concrete emerged. The presidencies of Portugal, Slovenia, France and the Czech Republic made no further progress on the issue.

Pilar Cancela, Secretary of State for International Cooperation of the Spanish government said, regarding the efforts that her government was making to hold the EU–CELAC Summit during the Spanish presidency: "Latin America is also a part of the international chessboard on which geopolitical rivalries are being played out. It has enormous potential and I think that Spain and Europe have to be in Latin America. This bi-regional relationship has to strengthen itself more in 2023 faced with the bloc of the United States, China or Russia" (Cancela, 2022). These arguments seem to reflect a common opinion in the EU institutions and the EU Member States, and which confirms our theories of the importance of the role that EU–CELAC inter-regionalism can play within the configuration of a new world order future.

The tension relations between the USA and Venezuela have also eased in recent months. The need for oil and gas supplies has been perhaps the main argument for re-establishing relations with Venezuela and therefore overcoming the disagreements between Brussels and Caracas, and maybe will resolve the political differences between EU and CELAC. Pragmatism in international relations can sometimes overcome concerns about principles and values. If the political crisis in Venezuela was the reason why relations between EU and CELAC were suspended, it should be remembered that in January 2021, with the constitution of the new Assembly of Venezuela, the EU was left without an institutional base to support Juan Guaidó as "President in charge". However, since that date the European Union has called Guaidó the "leader of the opposition" (El País, 2020).

China's increasingly strong position on the international stage was also a major issue that the Europeans were unable to resolve. While it is true that the EU and China have good trade relations, political relations are not so easy, especially in terms of human rights. It is not easy for EU Member States to deepen relations with China when they have to maintain relations with the USA. The pandemic caused by COVID-19 and recently Russia's invasion of Ukraine has only exacerbated an internal crisis in the EU and affected its foreign relations. All these issues have prevented a greater rapprochement with Latin America and the Caribbean. In addition, as has also been mentioned, most Latin American and Caribbean governments have chosen to deepen their relations with the Asia-Pacific region, and specifically with China. Russia's rapprochement with Latin America and the Caribbean is also a variable to consider for the European Union. There are currently several LAC countries that have close relations with Moscow (Gurganus, 2018). A recent example is Argentina's application to join the BRICS group of countries. The Chinese government supported this initiative (Ministry of Foreign Affairs, International Trade and Worship of Argentina, 2022a).

Although Russia has less influence than China in the Latin American and Caribbean region, its strategy of supplying its Sputnik vaccines has favoured it in strengthening its relations with several countries in the region. It should also be noted that in ideological terms many leftist parties took as a reference the communism promoted by the former USSR, and although that is past, the truth is that ideologically the Latin American left feels close to the ideas defended by Russia instead of from China. This historical aspect can be decisive in the foreign relations of several Latin American and Caribbean countries. Russia's rapprochement with governments such as Venezuela, Nicaragua, Cuba, Brazil and Argentina are clear examples of the influence that Moscow wishes to have within the region.

Although it seems that the new world order will be defined by countries' economic and commercial weight, undoubtedly ideologies will play an important role when countries decide to make alliances with one or another country. That is why the American, French, German, Chinese and Russian initiatives: The "Alliance for Multilateralism", "The Summit for Democracy", "Eurasian Economic Union" and the "Belt and Road Initiative" are initiatives that will be valued for their ideological component. The governments depending on political colour, will decide in which initiatives to participate and in which not. Since the Latin American and Caribbean countries do not have a common position on most foreign policy issues, we will find a fragmented Latin America and the Caribbean, an aspect that could hinder the USA, French, German, Russian and Chinese plans.

Another country that we have to take into account in EU–CELAC relations is India. India has a leading role in the Western strategy to consolidate the Indo-Pacific region, serving as a counterweight to China in the Asia-Pacific region. According to the Inter-American Development Bank "Latin America and India are not yet on each other's radar screen and neither can afford it. The center of gravity of the world economy has moved eastward and the region's integration with Asia has reached unprecedented levels. [LAC and India can cooperate] in education, science and technology and innovation, as it would facilitate the transfer of knowledge in renewable energy, information and communication technologies, healthcare, pharmaceuticals, higher education, mechanization, and new technologies applied to agroindustry, animal and plant health, biotechnology, technological development for livestock production, ocean research, among other" areas (Inter-American Development Bank and Export-Import Bank of India, 2019).

The Argentina PPT CELAC is working to relaunch CELAC-India relations. On 26 April 2022, the Minister of Foreign Affairs of Argentina, Santiago Cafiero, met with the Prime Minister of India, Narendra Modi. He expressed the intention of developing a work agenda that includes issues such as promoting an agenda for the Global South and the importance of coordinating in multilateral forums such as the World Trade Organization and the G20, based on the needs of emerging economies, highlighting that future summits of the G20 will take place in Indonesia, India and Brazil (Ministry of Foreign Affairs, International Trade and Worship of Argentina, 2022b).

Most of the leftist governments in Latin America and the Caribbean prefer to opt for a position of non-alignment between West and East. And as has been indicated, there is currently a preference for the Asia-Pacific region instead of preferring to strengthen the Atlantic axis. It remains to be seen if the government of President

Biden changes that perception of Latin Americans and Caribbeans for the Western world and they end up joining and defending it as their own model, and stop seeing the West as if it were a foreign space to them. In short, Latin America and the Caribbean are the West.

History shows us that the West was built taking as a reference the wisdom and technological advances developed in the East. We see how China and Russia, although confronting the West, have built a large part of their societies taking the West as a model. Therefore, East and West are not so far apart. What are far apart are their visions of democracy. At the end of the day, it seems that the capitalist model is not discussed by anyone, because everyone wants to implement it in their countries, starting with China itself, which has helped it become the world power it is today. Therefore, EU–CELAC inter-regionalism must, on the one hand, seek to defend the values that have made its civilisation, Western civilisation, a reference model for the rest of the world, through the promotion of universal values such as human rights, democracy, the rule of law and also seek to create a new inter-regionalism 4.0 that includes all the countries of both regions to create international standards that allow cyberspace to be regulated, with techplomacy being the foreign policy instrument to be implemented.

## References


Bateman, K. (2022). *Which is the world's most innovative cities in 2022?* World Economic Forum. Retrieved May 12, 2022, from https://www.weforum.org/agenda/2022/02/innovative-global-cities-talent-property/

Budapest Convention. (2001). *Convention on Cybercrime.* Budapest, 23, November 2001. Council of Europe. European Treaty Series N 185. Retrieved May 12, 2022, from https://rm.coe.int/1680081561

Cancela, P. (2022). *España presidirá la UE en 2023 centrada en la cooperación con América Latina.* Periódico Digital. Retrieved August 23, 2022, from https://newsinamerica.com/pdcc/boletin/2022/espana-presidira-la-ue-en-2023-centrada-en-la-cooperacion-con-america-latina/

CICTE. (2021). *OAS cybersecurity symposium.* September 20 to 30 2021. Retrieved May 12, 2022, from https://www.oas.org/en/sms/cicte/docs/AGENDA-SIMPOSIO-2021.pdf

Collier.P. (2018). *The future of capitalism: Facing the new anxieties.* Penguin Radom House.

Council of Europe. (2021). *Acceding to the Budapest convention on cybercrime: Benefits.* 4 June 2021. Retrieved May 12, 2022, from https://rm.coe.int/1680081561

Crafts, N. (2000). Globalization and growth in the twentieth century. *IMF working paper WP/00/44.* Research Department, International Monetary Fund. Retrieved August 23, 2022, from https://www.imf.org/external/pubs/ft/wp/2000/wp0044.pdf

Delhi Communiqué on a GFCE Global Agenda for Cyber Capacity Building. (2017). *Global forum on cyber expertise.* 24 November 2017. Retrieved May 12, 2022, from https://thegfce.org/wp-content/uploads/2020/04/DelhiCommunique.pdf

ECLAC. (2021). *Digital technologies for a new future.* Santiago: Economic Commission for Latin America and the Caribbean. Retrieved May 12, 2022, from Digital technologies for a new future (cepal.org).

El País. (2020). *La UE deja de reconocer a Guaidó como presidente pese a constatar su liderazgo.* Retrieved August 23, 2022, from https://elpais.com/internacional/2021-01-06/la-ue-reconoce-el-liderazgo-de-juan-guaido-pero-evita-reconocerlo-como-presidente.html

European Commission. (2017). *Joint communication to the European Parliament and the council. Resilience, Deterrence and Defence: Building strong cybersecurity for the EU*. Retrieved May 12, 2022, from https://eur-lex.europa.eu/legal-content/EN/TXT/PDF/?uri=CELEX:52017JC0450&from=en

European Commission. (2020). *New EU Cybersecurity Strategy and new rules to make physical and digital critical entities more resilient*. Retrieved May 12, 2022, from https://ec.europa.eu/commission/presscorner/detail/en/ip_20_2391

European Commission. (2020b). *Proposal for a regulation of the European Parliament and of the Council on contestable and fair markets in the digital sector Digital Markets Act*. Brussels, 15.12.2020COM(2020)842final.RetrievedAugust23,2022,fromhttps://ec.europa.eu/info/sites/default/files/proposal-regulation-single-market-digital-services-digital-services-act_en.pdf

European Commission. (2021). *EU-CELAC 2021–2023 Strategic Roadmap for the implementation of the Brussels Declaration and EU-CELAC Action Plan on Science, Technology, and Innovation*. Retrieved May 12, 2022, from https://ec.europa.eu/info/sites/default/files/research_and_innovation/strategy_on_research_and_innovation/documents/eu-celac_strategic-roadmap-2021-2023.pdf

European Commission. (2022a). *EU-US Trade and Technology Council* Retrieved May 16, 2022, from https://ec.europa.eu/info/strategy/priorities-2019-2024/stronger-europe-world/eu-us-trade-and-technology-council_en#background

European Commission. (2022b). *EU-US Trade and Technology Council: strengthening our renewed partnership in turbulent times*. Retrieved May 16, 2022, from https://ec.europa.eu/commission/presscorner/detail/en/IP_22_3034

European Council. (2021). *EU-Latin America & Caribbean Leaders' meeting: Join forces for a sustainable post COVID-19 recovery - press release by presidents Michel and von der Leyen*. European council. Retrieved May 12, 2022, from https://ec.europa.eu/commission/presscorner/detail/en/IP_21_6541

European External Action Service. (2021). *EU-LAC Partnership*. Digital Economy and Society. Retrieved May 12, 2022, from factsheet_eu_lac_en_29nov2021.pdf (europa.eu)

European Parliament. (2018). *Results of vote in Parliament*. Legislative Observatory, European Parliament, 12 March, 2019. Retrieved May 12, 2022, from 2017/0225(COD) - A8-0264/2018 (europa.eu)

European Parliament and Council. (2019). *Regulation (EU) 2019/881 of the European Parliament and of the Council of 17 April 2'19 on ENISA (the European Union Agency for Cybersecurity) and on information and communications technology cybersecurity certification and repealing Regulation (EU) N 526/2013* (Cybersecurity Act). EUR-Lex. Retrieved May 12, 2022, from EUR-Lex - 32019R0881 - EN - EUR-Lex (europa.eu)

Federal Council of Switzerland. (2020). *Digital Foreign Policy Strategy 2021–24*. Retrieved May 12, 2022, from https://www.eda.admin.ch/dam/eda/en/documents/publications/SchweizerischeAussenpolitik/20201104-strategie-digitalaussenpolitik_EN.pdf

Federal Department of Foreign Affairs. (2021). *Eleven norms of responsible state behavior in cyberspace*. Switzerland Confederation. Retrieved May 12, 2022, from https://www.eda.admin.ch/eda/en/fdfa/fdfa/aktuell/newsuebersicht/2021/04/uno-cyber-normen.html

Federal Foreign Office of Germany. (2019, May 29). *Future Affairs Berlin 2019*. Retrieved May 12, 2022, from http://futureaffairs19.re-publica.com/de/user/22022

Gang, Q. (2021). Dialogue of the heads of delegation of the member states, the secretary general and the assistant secretary general with permanent observers. Organization of American States. Retrieved May 12, 2022, from http://scm.oas.org/pdfs/2021/AG08444T.pdf

GFCE. (2015). *Launch of the global forum on cyber expertise*. Framework document. Retrieved May 12, 2022, from https://thegfce.org/wp-content/uploads/2020/04/framework-document-gfce.pdf

Gurganus, J. (2018). Russia: Playing a geopolitical game in Latin America. *Carnegie Endowment International Peace*, May 3, 2018. Retrieved August 23, 2022, from https://carnegieendowment.org/2018/05/03/russia-playing-geopolitical-game-in-latin-america-pub-76228

Horejsova, T., Ittelson, P., & Kurbalija, J. (2018). *The Rise of TechPlomacy in the Bay Area.* DiploFoundation (p. 12). Retrieved May 12, 2022, from https://www.diplomacy.edu/resource/the-rise-of-techplomacy-in-the-bay-area/#:~:text=Tech%20diplomacy%20is%20becoming%20a,to%20use%20innovative%20diplomatic%20approaches

Inter-American Bank and Export-Import Bank of India. (2019). *A bridge between India and Latin America. Policy options for deeper economic cooperation.* Retrieved August 23, 2022, from https://publications.iadb.org/publications/english/document/A_Bridge_Between_India_and_Latin_America_Policy_Options_for_Deeper_Economic_Cooperation_en_en.pdf

Kofod, J. (2021). The Techplomacy approach. *Ministry of Foreign Affairs of Denmark.* Office of Denmark's Tech Ambassador. Retrieved May 12, 2022, from https://techamb.um.dk/the-techplomacy-approach

Kurbalija, J. (2021). *2021: The emergence of digital foreign policy.* Diplo, 26 January 202. Recu Retrieved May 12, 2022, from https://www.diplomacy.edu/blog/2021-emergence-digital-foreign-policy

Maciel, M., Foditsch, N., Belli, & Castellon. (2016). *Cyber security, privacy and trust. Trends in Latin America* (pp. 7–12). Retrieved May 12, 2022, from ibliotecadigital.fgv.br/dspace/bitstream/handle/10438/16706/Cyber-Security%2C Privacy and Trust Trends in the Latin American and Caribbean region and the way forward.pdf

McKinsey Global Institute. (2019). Globalization in transition: The future of trade and value chains. *McKinsey Global Institute Report.* Retrieved August 23, 2022, from https://www.mckinsey.com/featured-insights/innovation-and-growth/globalization-in-transition-the-future-of-trade-and-value-chains

Ministère de L'Europe et des Affaires Étrangéres. (décembre 2017). *Stratégie internationale de la France pour le numérique.* Retrieved May 12, 2022, from https://www.diplomatie.gouv.fr/IMG/pdf/strategie_numerique_a4_02_interactif_cle445a6a.pdf

Ministry of Foreign Affairs of Denmark. (2017). *Foreign and security policy strategy 2017–2018.* Ministry of Foreign Affairs.

Ministry of Foreign Affairs of Denmark. (2021). *Strategy Denmark's Tech Diplomacy 2021–2023.*

Ministry of Foreign Affairs of Denmark. (2022). *How is Techplomacy operationalized.* Office of Denmark's Tech Ambassador. Retrieved May 12, 2022, from https://techamb.um.dk/the-techplomacy-approach

Ministry of Foreign Affairs of the Republic of Estonia. (2018). *Estonia appoints Heli Tiirmaa-Klaar as its first Ambassador at Large for Cyber Security.* Retrieved May 12, 2022, from https://vm.ee/en/news/estonia-appoints-heli-tiirmaa-klaar-its-first-ambassador-large-cyber-security

MinistryofForeignAffairs,InternationalTradeandWorshipofArgentina.(2022a).*Argentinaconsiguió el respaldo de China para ingresar a los BRICS.* Ministerio de Relaciones Exteriores, Comercio International y Culto de Argentina. Retrieved August 23, 2022, from https://www.cancilleria.gob.ar/es/actualidad/noticias/argentina-consiguio-el-respaldo-de-china-para-ingresar-los-brics

Ministry of Foreign Affairs, International Trade and Worship of Argentina. (2022b). En Nueva Delhi, Cafiero se reunió con el primer ministro Modi y coincidieron en la "importancia estratégica" del vínculo entre América Latina e India. Ministerio de Relaciones Exteriores. Comercio Internacional y Culto de Argentina. Retrieved August 23, 2022, from https://cancilleria.gob.ar/es/actualidad/noticias/en-nueva-delhi-cafiero-se-reunio-con-el-primer-ministro-modi-y-coincidieron-en

Niño, J. (2021). Dialogue of the heads of delegation of the member states, the secretary general and the assistant secretary general with permanent observers. *Organization of American States.* Retrieved May 12, 2022, from http://scm.oas.org/pdfs/2021/AG08444T.pdf

Norris, P., & Inglehart, R. (2019). *Cultural backlash: Trump, Brexit and authoritarian populism.* Cambridge University Press.

OAS and IDB. (2016). *Cybersecurity. Are we ready in Latin America and the Caribbean?* (p. 4). Organization of American States and Inter-American Development Bank.

OCDE. (2020). *Digital transformation in the age of COVID-19: Building resilience and bridging divides*. Digital economy outlook 2020 supplement. OCDE. Retrieved August 23, 2022, from https://www.oecd.org/digital/digital-economy-outlook-covid.pdf

Rodrik, D. (2019). Many forms of populism. VoxEU.org, 29 October 2019.

Schwab, K. (2015). The fourth industrial revolution. What it means and how to respond. *Foreign Affairs*. Council on Foreign Relations.

Schwab, K. (2016). *The fourth industrial revolution*. World Economic Forum.

SELA. (2022). *Specialization Course on Cyber Diplomacy and Specialization Course on Techplomacy.* Retrieved May 12, 2022, from http://www.sela.org/en/events/e/70982/cyber-diplomacy

Streeten, P. (2001). Integration, Interdependence, and Globalization. Finance & Development. A quarterly magazine of the *International Monetary Fund, 38*, 2. Retrieved August 23, 2022, from https://www.imf.org/external/pubs/ft/fandd/2001/06/streeten.htm

The EU Cybersecurity Act. (2019). *EUR-Lex.* Access to European law. Retrieved May 12, 2022, from https://eur-lex.europa.eu/eli/reg/2019/881/oj

The European Sting. (2022). *Why the European Union is opening a Silicon Valley "embassy"*. Retrieved August 18, 2022, from https://europeansting.com/2022/08/18/why-the-european-union-is-opening-a-silicon-valley-embassy/

The Hague Declaration on the GFCE. (2015). *The Hague: GFCE Secretariat*. Retrieved May 12, 2022, from https://thegfce.org/wp-content/uploads/2020/04/the-hague-declaration-on-the-gfce.pdf

Thoelken, H. (2019). *How can the foreign office remain an effective foreign policy player in the digital world of tomorrow? Geopolitics, future affairs*. Federal Foreign Office of Germany.

Torres Jarrín, M., & Riordan, S. (2019). Techplomacy. Hacia la búsqueda de una regulación del ciberespacio y la gobernanza de Internet. In B. de Moura (Ed.), *O Direito Internacional Privado Europeu: entre a harmonização e a fragmentação*. Emais Editora y Librería Jurídica.

Van Raemdonck, N. (2020). Balancing between giants, Latin-America's international cybersecurity position. *Elcano Royal Institute. ARI 104/2020.* Retrieved May 12, 2022, from https://media.realinstitutoelcano.org/wp-content/uploads/2021/10/ari104-2020-raemdonck-balancing-between-giants-latin-americas-international-cybersecurity-position.pdf

World Economic Forum. (2022). *Why the European Union is opening a Silicon Valley "embassy"*. Retrieved August 16, 2022, from https://www.weforum.org/agenda/2022/08/why-the-european-union-is-opening-a-silicon-valley-embassy/

# Chapter 10
# Conclusions

The twenty-first century has become a period of constant crises. The SARS epidemic was followed by the Great Financial Crisis, followed by the COVID-19 Pandemic, followed by the Russian invasion of Ukraine. The 1990s were a decade of economic and political convergence when it was believed that disparate governance models would slowly converge on the Western liberal market economy. Western foreign ministries and militaries were reconfigured to manage the existential challenges of the New International Security Agenda: international terrorism, climate change, migration and pandemic disease. As Fukuyama wrote, "The End of History". But geopolitics is back with a vengeance as contests over territory, resources, the balance of power and spheres of influence undermine our capacity to manage the big global issues. All this has been played out against the background of advances in innovation generated by Big Tech Companies such as cyberspace, AI, Internet of Things, Machine Learning and Quantum Computing. These technologies have revolutionised the way companies produce goods and services and the way in which people, businesses, institutions and governments communicate with each other.

Information is a crucial element in the daily work of foreign ministries and their diplomatic services and how they promote and protect their country's national interests. Digital technologies long ago radically changed the ways in which diplomats can gather and communicate information. Social media platforms have become a central part of public diplomacy communicating with foreign publics but also advising their citizens in consular emergencies. The more innovative foreign ministries are experimenting with computer modelling and big data analysis. COVID-19 broke the great taboo and forced diplomats to negotiate and hold meetings online. But these technologies on which diplomats are becoming increasingly dependent are vulnerable to cyberattack and infiltration. They are produced by technology and Internet companies often larger than most countries and over whom governments have only limited control.



M. Torres Jarrín, S. Riordan, *Science Diplomacy, Cyberdiplomacy and Techplomacy in EU-LAC Relations*, United Nations University Series on Regionalism 25, https://doi.org/10.1007/978-3-031-36868-4_10

Science Diplomacy has long sought to promote international scientific exchanges and use scientific exchanges themselves to promote international harmony. The COVID-19 pandemic was a great test of its effectiveness. The science did well, the diplomacy less so. The geopolitical blame game undermined international solidarity and collaboration. Countries exchanged accusations and insults about the origins of the disease with little regard for the science. In the early stages of the pandemic governments fought each other over protective equipment. Western governments prioritised booster jabs of their own populations over sharing vaccine doses more equitably allowing China and Russia to move into the gap with their "vaccine diplomacy". The big pharmaceutical companies, who developed the vaccines on which many countries' anti-COVID-19 strategies depended, reinforced their role as major international players but many governments had problems in managing their relationships with them. Techplomacy is not limited only to digital technology and internet companies. COVID-19 will not be the last pandemic to strike mankind this century, and Science Diplomacy as a discipline needs to think hard about why it delivered so little this time and how it can do better in the future.

For too long cyberspace was left to the technicians. It was seen as a technological construction whose problems were essentially technical and could be left to the Internet companies and their engineers. The problems companies like Meta/Facebook have encountered designing algorithms to police unacceptable content on their platforms show this is not so. Indeed, cyberspace generates a whole host of political and geopolitical problems centred on both Internet governance and cybersecurity which require diplomacy at an international level. Although we think of cyberspace as something virtual floating in the cloud, the internet is physically embedded in continental and oceanic Internet cables and switching stations and data storage facilities. All are vulnerable to physical as well as cyberattack. If, as Clemenceau once said, war is too important to leave to generals, cyberspace is definitely too important to leave to technicians. Diplomats need to combine better expert knowledge with their diplomat skills and techniques to begin building rules of the game for cyberspace, adapting where necessary the provisions of international law to the realities of this new domain. If the international community of diplomats is a necessary adjunct to international stability, so an international community of cyberdiplomats will be necessary to generating and maintaining an orderly cyberspace.

But managing security and governance in cyberspace does not depend just on governments and their diplomats. As this book has repeatedly made clear, the major technology and internet companies are now geopolitical actors in their own right. Not only are they among the largest and wealthiest companies in the world, with market capitalisations exceeding the GDP of all but the largest countries, but they also develop and engineer the digital and other technologies on which modern societies, and international society, depend. Companies that propose developing their own digital currencies or constructing global virtual worlds are behaving as geopolitical actors even if they are reluctant to admit it to themselves. Any process to regulate cyberspace must engage with these companies. But governments have been slow, or reluctant, to do so. The European Commission seems almost to prefer to regulate against US internet companies, not regarding them as part of the process (this may reflect the lack of major internet companies in the EU). But this is unrealistic. The technology and Internet companies have created the domain which governments now seek to regulate. They must form part of the regulatory process alongside other interested non-state actors like NGOs and user groups. Cyberdiplomacy must be multistakeholder diplomacy.

But Techplomacy goes beyond involving technology and Internet companies in debates about the regulation of cyberspace as just another non-governmental stakeholder. In techplomacy, a government engages with these companies as it would with other governments. It seeks information about the relations between the companies and new innovations and technological developments. It promotes itself as a location for research and production centres. It debates international regulatory issues and tries to convince them to support its preferred solutions. It seeks to engage them in coalitions of the willing. It looks to protect its citizens rights in relation to issues like privacy and protection of data. In short, it seeks to replicate the political, economic, commercial and consular roles of an embassy in a country. That is why Denmark chose to nominate an ambassador to the tech sector and not another digital envoy. Therefore, his job description reads like that of an ordinary ambassador to a country. The corollary is that if governments need to develop a techplomacy to engage with the major technology and internet companies then these companies need to develop their own techplomacy for engaging with governments. Microsoft at least seems to have recognised this, head-hunting Denmark's first Tech ambassador to head up Microsoft's public affairs operation in Europe—in effect becoming Microsoft's ambassador to the EU.

The strategic partnership between EU and CELAC needs to be renewed. A plan to resurrect the EU-CELAC summits has already borne fruit. In their absence, the EU has lost ground in Latin America. Countries like Russia, China and even Saudi Arabia have increased their economic presence and political influence in the region. Argentina has applied to join the BRICS. Brazil may follow suit. Of particular concern for the EU should be the growing economic and political influence of China. The EU's limited economic resources make it hard for it to compete with China's Belt Road Initiative, especially with the EU being forced to focus on post-COVID recovery and the possible reconstruction of Ukraine. Cooperation in the areas of science diplomacy, cyberdiplomacy and techplomacy offer promising new paths for EU-CELAC cooperation. The two regions regions have both the disadvantage and the advantage of not having really significant technology and internet companies of their own. This is a disadvantage because it threatens to reduce their influence in the key debates in which these companies will participate. It is an advantage as neither region is beholden to the interests or demands of such companies. Both regions find themselves balancing between the technological giants of the USA and China. Their own digital and AI research projects are limited. Working together to develop joint positions, as in the past on other issues, they can both multiply their influence in key debates about international cyber and technology norms while occupying a neutral ground between the USA and China. There is much that CELAC countries can learn from the EU's regulatory approach to new technologies, including the protection of citizens' data. Close collaboration between EU and CELAC on cyber-related issues may also help limit or manage China's digital penetration of LAC countries, which threatens their freedom of geopolitical manoeuvre and the scope of future EU-CELAC cooperation, without provoking the confrontation with Beijing of the USA's more aggressive approach. There is a threat of a growing technology gap between countries like China and the USA and the rest of the world. This would have economic as well as geopolitical consequences.

It would be too much to claim that cooperation over science, cyber and technology diplomacy is a panacea for the restoration of the EU's relations with Latin America/Caribbean. Historically, the EU has not taken these countries as seriously as it should while being frustrated by the inability of the LAC countries to develop effective regional organisations of their own. But science, cyber and technology do offer an area where the EU and CELAC can collaborate to mutual benefit in areas key to the future economic and political development of both regions. Ultimately, EU-CELAC inter-regionalism in these areas could create a new inter-regionalism 4.0 for a new global governance 4.0.

# Correction to: Science Diplomacy, Cyberdiplomacy and Techplomacy in EU-LAC Relations

**Correction to:**
**M. Torres Jarrín, S. Riordan, *Science Diplomacy, Cyberdiplomacy and Techplomacy in EU-LAC Relations*,**
**United Nations University Series on Regionalism 25**
**https://doi.org/10.1007/978-3-031-36868-4**

The book was inadvertently published with an incorrect Volume number 3. This has now been amended throughout the book to the correct Volume number 25.

---

The updated version of the book can be found at
https://doi.org/10.1007/978-3-031-36868-4



M. Torres Jarrín, S. Riordan, *Science Diplomacy, Cyberdiplomacy and Techplomacy in EU-LAC Relations*, United Nations University Series on Regionalism 25, https://doi.org/10.1007/978-3-031-36868-4_11

MIX
Papier aus verantwortungsvollen Quellen
Paper from responsible sources
FSC® C105338

If you have any concerns about our products,
you can contact us on
**ProductSafety@springernature.com**

In case Publisher is established outside the EU,
the EU authorized representative is:
**Springer Nature Customer Service Center GmbH**
**Europaplatz 3, 69115 Heidelberg, Germany**

Printed by Libri Plureos GmbH
in Hamburg, Germany